Darwinism and Pragmatism

Charles Darwin's theory of natural selection challenges our very sense of belonging in the world. Unlike prior evolutionary theories, Darwinism construes species as mutable historical products of a blind process that serve no inherent purpose. It also represents a distinctly modern kind of fallible science that relies on statistical evidence and is not verifiable by simple laboratory experiments. What are human purpose and knowledge if humanity has no pre-given essence and science itself is our finite and fallible product?

According to the Received Image of Darwinism, Darwin's theory signals the triumph of mechanism and reductionism in all science. On this view, the individual virtually disappears at the intersection of (internal) genes and (external) environment. In contrast, William James creatively employs Darwinian concepts to support his core conviction that both knowledge and reality are in the making, with individuals as active participants. In promoting this Pragmatic Image of Darwinism, McGranahan provides a novel reading of James as a philosopher of self-transformation. Like his contemporary Nietzsche, James is concerned first and foremost with the structure and dynamics of the finite purposive individual.

This timely volume is suitable for advanced undergraduate, postgraduate and postdoctoral researchers interested in the fields of history of philosophy, history and philosophy of science, history of psychology, American pragmatism and Darwinism.

Dr Lucas McGranahan is an independent scholar with research interests in American pragmatism, nineteenth-century philosophy and the history and philosophy of science.

T0266482

History and Philosophy of Biology

Series editor: Rasmus Grønfeldt Winther | rgw@ucsc.edu | www.rgwinther.com

This series explores significant developments in the life sciences from historical and philosophical perspectives. Historical episodes include Aristotelian biology, Greek and Islamic biology and medicine, Renaissance biology, natural history, Darwinian evolution, nineteenth-century physiology and cell theory, twentieth-century genetics, ecology, and systematics, and the biological theories and practices of non-Western perspectives. Philosophical topics include individuality, reductionism and holism, fitness, levels of selection, mechanism and teleology, and the nature–nurture debates, as well as explanation, confirmation, inference, experiment, scientific practice, and models and theories vis-à-vis the biological sciences.

Authors are also invited to inquire into the 'and' of this series. How has, does, and will the history of biology impact philosophical understandings of life? How can philosophy help us analyse the historical contingency of, and structural constraints on, scientific knowledge about biological processes and systems? In probing the interweaving of history and philosophy of biology, scholarly investigation could usefully turn to values, power, and potential future uses and abuses of biological knowledge.

The scientific scope of the series includes evolutionary theory, environmental sciences, genomics, molecular biology, systems biology, biotechnology, biomedicine, race and ethnicity, and sex and gender. These areas of the biological sciences are not silos, and tracking their impact on other sciences such as psychology, economics, and sociology, and the behavioural and human sciences more generally, is also within the purview of this series.

Rasmus Grønfeldt Winther is Associate Professor of Philosophy at the University of California, Santa Cruz (UCSC), and Visiting Scholar of Philosophy at Stanford University (2015–2016). He works in the philosophy of science and philosophy of biology and has strong interests in metaphysics, epistemology, and political philosophy, in addition to cartography and GIS, cosmology and particle physics, psychological and cognitive science, and science in general. Recent publications include The Structure of Scientific Theories in *The Stanford Encyclopaedia of Philosophy*; and Race and Biology in *The Routledge Companion to the Philosophy of Race*. His book with University of Chicago Press, *When Maps Become the World*, is forthcoming.

Darwinism and Pragmatism

William James on evolution and
self-transformation

Lucas McGranahan

Routledge
Taylor & Francis Group

LONDON AND NEW YORK

First published 2017 by Routledge

2 Park Square, Milton Park, Abingdon, Oxfordshire OX14 4RN
52 Vanderbilt Avenue, New York, NY 10017

Routledge is an imprint of the Taylor & Francis Group, an informa business

First issued in paperback 2019

British Library Cataloguing in Publication Data
A catalogue record for this book is available from the British Library

Library of Congress Cataloging in Publication Data
A catalog record for this book has been requested

ISBN: 978-1-8489-3596-9 (hbk)
ISBN: 978-0-367-35857-0 (pbk)

Typeset in Times New Roman
by Wearset Ltd, Boldon, Tyne and Wear

For my mother

Contents

Acknowledgements

This book would not exist without Rasmus Grønfeldt Winther. It was Rasmus who first suggested that I examine pragmatism through Darwinism during my PhD studies at the University of California, Santa Cruz. He is an invaluable guide through the history and philosophy of science, as well as a friend. *Tak, gracias*, and thank you.

I am also indebted to David Hoy, who drew me to Santa Cruz in the first place. David greatly expanded my knowledge of European philosophy, and his writing is a model of clarity. I am privileged to know someone so erudite and yet so genuine and kind. Jocelyn Hoy was similarly welcoming as well as an excellent teacher; assisting her in teaching Nietzsche to undergraduates was a particular pleasure. Ellen Suckiel's careful work on William James – including her rigorous seminar on the topic – gave me the confidence to make James the focus of my studies. Colin Koopman was also an asset, especially because of our shared interest in connecting American and Continental thought. In addition, many of the ideas in this book were first workshopped by fellow graduate students Jacob Metcalf, Kaija Mortensen, Benjamin Roome and Andrew Delunas. Such a 'Metaphysical Club' provides a kind of intellectual camaraderie that comes only from one's peers.

Thank you also to the Society for the Advancement of American Philosophy (SAAP), which shows by example that philosophers can be practical, cooperative and socially engaged. My scholarship was honed at SAAP meetings in conversation with Trevor Pearce, Michael Brady, Mark Moller, Megan Mustain, Nick Smaligo and Randall Auxier, among others. SAAP also encouraged this project in its early stages by printing my article 'William James's Social Evolutionism in Focus' in its journal *The Pluralist* (issue 6, no. 3), published by University of Illinois Press. Chapter 2 is a downstream version of the latter article.

A self-portrait by William James is included by permission of the Houghton Library, Harvard University (William James Drawings, 1859–1880, MS Am 1092.2, 54).

The above debts, while significant, were mostly incurred during an earlier, more academic phase of my life. Support today looks less like a conference or writing group and more like the increased freedom and energy to express myself through theatre, music and a wider variety of writing styles. I would therefore like to thank all of my creative co-conspirators – especially those at Pan Theater in Oakland – for torqueing the cognitive-affective-aesthetic dialectic in which I consist. Who needs a hypertrophied intellect?

I am grateful to everyone, including those I have inadvertently omitted. I take responsibility for all errors and omissions. I deny, however, that avoiding error is our sole task in a world of finite time and finite rationality.

Abbreviations

The Works of William James. Edited by F. Burkhardt, F. Bowers and I. K. Skrupskelis. Cambridge, MA: Harvard University Press. 1975–1988. Original date of publication is given in parentheses.

ECR   *Essays, Comments, and Reviews*. 1987.
EPH   *Essays in Philosophy*. 1978.
EPS   *Essays in Psychology*. 1983.
ERE   *Essays in Radical Empiricism*. 1976 (1912, posthumous).
ERM   *Essays in Religion and Morality*. 1982.
MEN   *Manuscript Essays and Notes*. 1988.
ML   *Manuscript Lectures*. 1988.
MT   *The Meaning of Truth*. 1975 (1909).
P   *Pragmatism*. 1975 (1907).
PP   *The Principles of Psychology*. 2 vols. 1981 (1890).
PU   *A Pluralistic Universe*. 1977 (1908).
SPP   *Some Problems of Philosophy*. 1979 (1911, posthumous).
TT   *Talks to Teachers on Psychology: And to Students on Some of Life's Ideals*. 1983 (1899).
VRE   *The Varieties of Religious Experience: A Study in Human Nature*. 1985 (1902).
WB   *The Will to Believe and Other Essays in Popular Philosophy*. 1979 (1897).

Correspondence

CWJ   *The Correspondence of William James*. 12 vols. Edited by I. K. Skrupskelis and E. M. Berkeley. Charlottesville: University of Virginia Press. 1992–2004.
LWJ   *The Letters of William James: Two Volumes Combined*. Edited by H. James. Boston, MA: Little, Brown, and Co. 1920.

William James self-portrait. Pencil on paper, mid-1860s.

Introduction

A pragmatic image of Darwinism

Charles Darwin's epoch-making claim in *On the Origin of Species* (1859) was not that species change over time or give rise to other species. Such ideas had already been advanced by Enlightenment thinkers such as Erasmus Darwin and Jean-Baptiste Lamarck. Lamarck in particular had put forth a comprehensive evolutionary theory and claimed in his *Philosophie zoologique* (1809) that humans had descended from a kind of ape. What set Charles Darwin apart from earlier evolutionists was his distinctive mode of explanation. Darwin summarizes this idea as follows:

> Variations, however slight and from whatever cause proceeding, if they be in any degree profitable to the individuals of a species, in their infinitely complex relations to other organic beings and to their physical conditions of life, will tend to the preservation of such individuals, and will generally be inherited by the offspring. The offspring, also, will thus have a better chance of surviving, for, of the many individuals of a species which are periodically born, but a small number can survive. I have called this principle, by which each slight variation, if useful, is preserved, by the term Natural Selection.[1]

Assuming a 'struggle for existence' in which not all individuals can succeed equally, those whose traits increase their relative ability to survive and reproduce will tend on average to leave more offspring, thus spreading their (heritable) traits in the population. This process is meant to explain not only changes in a species, but also, given enough time and divergence of characteristics, the creation of new species themselves.

The challenge of Darwinism

The idea of natural selection may seem familiar or even obvious today. However, its implications remain largely undigested by both intellectual and popular culture. The problem is not just the humbling suggestion that humans are – in a literal genealogical sense – closely related to other primates. The challenging implications of natural selection go beyond this and are not necessarily shared by other evolutionary theories. Consider briefly four features of Darwin's theory.

First, *Darwinism does not grant humans an inherently special place in evolution*. Strictly speaking, this point is less about natural selection per se than it is about the genealogical pattern that natural selection is meant to predict and explain.[2] Whereas Enlightenment evolutionists had arguably just temporalized the Great Chain of Being by positing a linear progression of living beings crowned by the human species, Darwin proffered anatomical and geographical evidence to argue that evolution is in fact a nonlinear branching process. The pattern of evolution is not a march from less perfect to more perfect. It is more like Darwin's image of the ramifying 'tree of life' – and likely something even more gnarled and reticulated than this. Humans occupy one branch among others, and it is a contingent fact of history that this branch exists at all. We may be of particular interest (to ourselves) due to our capacities for reasoning, culture and language, but evolution does not lead to *Homo sapiens* any more than it leads anywhere else. As Freud has suggested, this decentring of humankind within the community of life may be viewed as a biological analogue of the Copernican decentring of humankind within the cosmos. That is to say, it has been a difficult thing for the ego to cope with.

Second, *natural selection is by definition a mindless and non-intentional process*. This does not mean that minds do not exist or that no one can intentionally affect evolutionary processes. Rather, it means that explanations in terms of natural selection do not invoke intentions, and certainly not in the form of an overall goal toward which the process is aiming. Evolution does not pulse toward some preordained conclusion, nor is an anthropomorphic Mother Nature picking out her favourite forms and culling the weaklings. She is only a metaphor, as indeed is 'selection' if this term is taken to imply intention. Indeed, this is Darwin's very point in choosing the term 'natural selection', as opposed to the *artificial* selection practiced by breeders.[3] Darwinism thus undermines the arguments of natural theology, which rely on the intuition that useful designs must have been designed intelligently. Insofar as natural selection succeeds in explaining the exceptional variety and adaptedness of organic forms, an intentional force is rendered superfluous for this same task. Why invoke purpose or intention where a mindless process will do?

Third, *natural selection undermines essentialism, while foregrounding relations and indeterminacy*. This point may seem more dryly technical than the first two, but it is in a way more fundamental. Nothing seems more obvious than that reality is composed of discrete objects that are defined by the categories to which they belong. The world is a world of categorized things. As John Dewey has argued,[4] this basic feature of our phenomenology props up a metaphysics that has held considerable sway over the philosophical tradition since Plato: Objects are real to the extent that they instantiate the categories to which they belong; reality is in the general and not in the particular. Such a view draws particular support from observations of the organic world. How would generations upon generations of oak trees – or ravens, or humans – demonstrate such consistent and characteristic life histories if not for some guiding form defining their reality? An ideal essence must shepherd the concrete individual into existence, and the individual must somehow fall short to the extent that it differs from this eternal form.

Darwinism suggests a different conception of form, reflecting what Ernst Mayr has called 'population thinking'. As opposed to the 'typological thinking' just described, population thinking construes the type as a statistical aggregation of information about individuals.[5] This is a full conceptual about-face: The species is now a dynamic abstraction gleaned continually after the fact, not a form-giving essence giving rise to approximations of itself. This mindset promotes a new conception of variation, conceived not as an array of monstrosities clustering around a fixed ideal, but as a shifting overall state of differences among individuals. Variation is not metaphysical failure or organic 'noise' but the precious raw material for evolutionary change. Theorists today may still argue about what constitutes the 'essence' of a species, but they do not typically posit essences in the ancient sense of idealized forms that exist independently of the natural world.[6] We are population thinkers now.

Fourth, *Darwinism represents a distinctly modern kind of fallible science.* Natural selection was among the first major scientific theories to rely heavily upon probabilistic evidence. It was thus a major player in the probabilistic revolution that swept through nineteenth-century science, reconfiguring traditional notions of evidence and proof.[7] Darwin could not demonstrate with certainty that natural selection has been a significant factor in evolution, nor could he show that any particular trait was certain to have arisen or will definitely arise in the future. He could only claim that with enough heritable variation – of the right kinds, with enough frequency – one could reasonably expect to see certain types of evolutionary change that we do in fact see. Darwin thus seems to have employed what his contemporary Charles Sanders Peirce termed *abduction*, better known as *inference to the best explanation*.[8] Abduction is not naively inductive or Baconian, as if one could read laws directly off observations of nature. Rather, it allows a role for creativity, analogy and intuition. Thus, as much as Darwin was a careful naturalist, natural selection could be described as a powerful guess – spurred by an analogy with selective breeding and a generalization of Thomas Malthus's social 'struggle for existence' – that made intelligible a set of disparate facts about anatomy and biogeography. Such abductive guessing is as much a matter of showing how disparate ideas hang together as of making specific testable predictions. This methodology was only beginning to be theorized in Darwin's time by thinkers like Peirce and William Stanley Jevons, however. To many it still looked suspiciously like putting the theoretical cart in front of the empirical horse. Darwin's epistemic problems were compounded by the speculative nature of his theory of heredity and by geologists' low estimates of the timespan in which evolution had to do its work. The eventual triumph of natural selection in biology was thus in no way obvious during Darwin's lifetime. Darwinism might have been acceptable to a reasonable modern fallibilist, but not to a scientific conservative whose only ideal of proof was the reproducible experiments of classical physics.

Taking these points together, it is clear that natural selection challenges some of our most entrenched habits of thinking as well as our very existential orientation in the world. Daniel Dennett has thus described Darwin's theory as a corrosive 'universal acid' that transforms everything in its wake.[9] What is our purpose if humanity is neither a pre-given goal nor the expression of a fixed essence? What happens when absolute cosmic and Platonic teleologies are unmasked to reveal only a swarm of particular relative teleologies – that is, nothing but the finite purposiveness of concrete living beings? What if knowledge itself is an evolved function of finite creatures that can only glimpse their cosmos and ancestry through biased myopic guesswork? This is a tangled nest of questions with no easy answer.

One response to this distinctively modern problematic is represented by the Received Image of Darwinism.[10] For the Received Image, Darwinism signals the triumph of mechanistic determinism in all of science. By replacing teleology with population thinking and blind selection, Darwinism gives universal scope to the methods of classical physics. Without cosmic or individual purposes to account for, all phenomena may now be explained in terms of mechanistic matter-in-motion. This opens the door to an image of natural selection as a great deterministic mill for the culling of variation.[11] Darwinism may posit 'random' variation and trade in probabilities, but none of this makes it incompatible with such determinism. (Darwinian variation is only 'random' in the sense of not being directed by environmental demands.) Add to this the gene-centric view popularized by Richard Dawkins in *The Selfish Gene* (1976), and the real protagonists in evolution are no longer organisms but the genes that construct them as their hapless vehicles.

The Received Image of Darwinism thus plays easily into a philosophy of 'nothing but'. Ultimately, we are nothing but a product of the long-term mechanical sorting of mutations by the environment. Proximally, we are nothing but genetic programmes being triggered by impersonal external factors. In any event, the individual organism – the actual concrete seat of life from which our biological questioning originates – has fallen out of favour as a level of analysis or locus of causation. Organisms are replaced by abstract gene pools; purpose is ignored instead of reconstructed; and Enlightenment ideals of rational mechanistic science are left intact.

There is an irony in the Received Image of Darwinism, which parallels an irony in the work of Freud (already mentioned above). Freud spent an entire career undermining traditional estimations of the scope and self-transparency of human rationality. Nevertheless, he saw himself as continuing the Enlightenment project of replacing unreason with rational mechanistic science. Unlike his psychoanalytic and literary followers, Freud was trained in reductionistic neurophysiology and never abandoned its methodological assumptions.[12] Freud's critique of rationality thus does not come full circle in comprising a critical reconstruction of his own aims and methods. Similarly, the Received Image of Darwinism assimilates the science of life to an Enlightenment model of physics that is no longer universally accepted even in physics. If historian of psychology Eugene Taylor is correct that no science is mature that has not attained self-awareness about its own epistemology,[13] then Darwinians might look to the resources of their own science in undertaking a genuine reconstruction of individuality, purpose and knowledge.

What if this occurred in the immediate wake of Darwin's *Origin* and we simply ignored it?

William James

This book sketches an alternative to the Received Image of Darwinism through an examination of the writings of seminal American thinker William James (1842–1910). James pioneered a Darwinian approach in both psychology and philosophy, prefiguring twentieth-century views in several regards. Interestingly, however, James uses Darwinian concepts precisely to underwrite his analysis of individual agency and his overall philosophical pluralism. James thus advances a Pragmatic Image of Darwinism that contrasts starkly with the Received Image outlined above.

James's Pragmatic Image of Darwinism is characterized by the following features:

- *Internalism and constructionism:*[14] The environment does not shape individuals or populations as passive clay. Both ontogeny and phylogeny are directed in part by processes that are internal to organisms (internalism). The individual organism is a locus of agency in its environment (constructionism).
- *Generalized selectionism*: Natural selection is one instance of a general pattern that exists in various domains and at various levels of analysis. This pattern may be reiterated at multiple hierarchical levels, where selection at one level provides variation for another. For instance, the human being is comprised by a hierarchical selectionist sensorimotor system that is mediated by a selective will.
- *Fallible knowledge and indeterminate truth*: Knowledge is fallible, partial and inductive, across all domains. Meaning is determined in an ongoing fashion in relation to changing environments. Beliefs become true by their workings in concrete experience. There is no standpoint from which the world appears as a single fact.
- *Dynamically continuous reality*: Reality is non-deterministic and includes genuine alternative possibilities. All forms, whether conceptual or material, are temporal and provisional. Relations are as ontologically primordial as the things or concepts they relate.

It is important to note that this worldview is not creationist or otherwise in denial of the four Darwinian challenges introduced above: James accepts the decentring of humankind in the tree of life; the lack of a pre-given purpose or direction in ontogeny and phylogeny; and the fallible and abductive nature of science. James's worldview is thus a *Darwinian* way of responding to Darwinism, not ostrich-in-the-sand behaviour.

To understand this position fully will require a careful exploration of the variegated terrain of nineteenth-century philosophy, psychology and biology. At the outset, however, it will be instructive to begin with a sketch of our protagonist himself.[15]

Centres of vision

This book adopts a hermeneutical principle from its subject of study. According to William James, 'Any author is easy if you can catch the centre of his vision'.[16] This means something more than a mere rational reconstruction of abstract claims and inferences. In James's words,

> Get at the expanding centre of a human character, the *élan vital* of a man, as Bergson calls it, by living sympathy, and at a stroke you see how it makes those who see it from without interpret it in such diverse ways.... Place yourself similarly at the centre of a man's philosophic vision and you understand at once all the different things it makes him write or say. But keep outside, use your post-mortem method, try to build the philosophy up and out of the single phrases, taking first one and then another and seeking to make them fit 'logically', and of course you fail. You crawl over the thing like a myopic ant over a building, tumbling into every microscopic crack or fissure, finding nothing but inconsistencies, and never suspecting that a centre exists.[17]

James's point is that interpretation is facilitated by entering into a state of holistic cognitive-affective empathy with the imagined author. Consistency in this undertaking is desired, but to focus entirely on comparing sentences is to miss the point. As James once noted in a letter to a critical student, 'This is splendid philology, but is it a live criticism of anyone's *Weltanschauung*?'[18]

The point here is not to project an unrealistic level of coherence on James as an author or human being. Michel Foucault may be correct that the author is a historically contingent functional principle rather than an originary source that antecedes the text.[19] However, this simply means that the author is like all concepts: a provisional construction that may be critiqued based on its role within discourse (or, in more Jamesian terms, *experience*). More interesting than discarding author-talk as meaningless is to consider its limitations and advantages in different contexts. For its part, this book wagers that both philosophy and science benefit from *idiographic* studies – examinations of individuals or particulars as such – in addition to studies of general laws, statistical norms, disembodied intellectual trends and the like. This includes idiographic studies of authors as exemplars of worldviews. Philosophies are lived by whole concrete individuals, and something is lost when they are removed from this ecological context. The purpose here is to reconstruct James's vision charitably, as a powerful worldview that he developed over the course of a lifetime. Such a worldview, surveyed in its entirety, may be evaluated as a unique response to scientific and philosophical problems still extant today. A philosophy developed in immediate response to Darwin may yet provide a useful corrective to a calci-fied neo-Darwinism.

To be fair, this methodology presumes something of what the book would like to promote: a Jamesian belief in the value of examining and emulating indi-viduals as such. Some such circularity is difficult to avoid, but it is no more offensive to logic than the presumption that individuals are negligible. In any event, studies making alternative wagers may be viewed as complementary rather than contradictory.

Life and works

William James was born into an affluent family in New York City in 1842. The James family moved often, both within the north east of the United States (US) and between this area and Europe. This gave the James children a culturally rich upbringing but a highly fragmented education. William James's father, Henry James, Sr., was a minor philosophical author and a follower of a mystical sect of Christianity known as Swedenborgianism. A cosmopolitan intellectual, the elder James kept illustrious company. Among the James family friends were Scottish historian and essayist Thomas Carlyle and New England transcendentalist Ralph Waldo Emerson, who seems to have been William's godfather.[20] Of course, the family was not just in good intellectual company but would produce two great luminaries of its own: Celebrated novelist Henry James (Jr.) was William's younger brother, and William himself practically founded scientific psychology in the US in addition to popularizing American pragmatism, the country's first philosophical school of international consequence.

William James's path to these ends was circuitous. As a teenager James studied to become a painter. However, he abandoned his artistic aspirations in 1861 and entered Harvard University's Lawrence Scientific School. The timing was fateful, as Darwin had just published the *Origin of Species* in 1859. At Harvard James had access to leading minds who were grappling with Darwin's theory, including anatomist Jeffries Wyman, botanist Asa Gray, anti-Darwinian zoologist Louis Agassiz, evolutionary philosopher John Fiske and positivist champion of Darwinism Chauncey Wright, among others. James began his studies at Harvard in chemistry but quickly switched to anatomy and physiology, which he used as a basis for studying medicine at Harvard Medical School beginning in 1864. After a couple of significant interruptions – an expedition to Brazil with Agassiz in 1865/1866 and a trip to Germany to study the new science of psychology in 1867/1868 – James finally took his medical degree in 1869. James would never practice medicine nor take any other earned degree. In James's words, 'I originally studied medicine in order to be a physiologist, but I drifted into psychology and philosophy from a sort of fatality'.[21]

The stamp of evolutionary debates is evident in James's very earliest writ-ings. James's first publications were two cautiously positive reviews of promi-nent evolutionists in 1865, one of 'Darwin's Bulldog' Thomas Henry Huxley, and another of natural selection's co-discoverer Alfred Russell Wallace. In 1868, James published two somewhat sceptical pieces on Darwin's *Variation of Animals and Plants under Domestication*, which lays out the author's speculative theory of variation and heredity. Subsequent publications show James repeatedly making a foil of English evolutionary philosopher Herbert Spencer, whom he had greatly admired in his youth. Spencer was the subject of James's first signed scholarly essay in 1878, a Darwin-inspired critique of evolutionary psychologies that ignore the role of individual biases and interests in the construction of knowledge.[22] An 1880 essay on social evolution continues the critique on a different level, using Darwinian concepts to argue for the importance of individuals in history.[23] In all these writings, James treats natural selection as a suggestive source of ideas, without demonstrating a dogmatic devotion to the particulars of Darwin's theory.

Although James liked to claim that philosophy is a function of temperament, his emphasis on individual agency and freedom was not grounded in a naturally sanguine outlook. The truth is the opposite: The possibility of freedom for James was a fallible expression of hope posited against the background of despair. Of particular note is a spate of severe depression that James suffered during the late 1860s and early 1870s. This depression was fuelled by his increasingly materialistic worldview, reinforced by his studies in the new field of physiological psychology. James was lifted from his depression by affirming a conception of freedom that he discovered in the writings of French neo-Kantian philosopher Charles Renouvier. Freedom on this view means simply the ability to attend to one idea among others, such that the idea's associated motor consequences are allowed to follow. James considers it to be possible (if not provable) that such attending may occur freely, in a non-predetermined fashion. Given that freedom is possible, James chooses to act as if it is real. James's philosophy is built around this auto-affirmation of freedom rather than a purported foundational truth. This provides a deep sense in which it is practical or pragmatic rather than rationalistic.

James was also a beloved teacher with a long academic career. He was hired by his alma mater Harvard as Instructor in Physiology in 1872, where he would remain under various titles until 1907. As a Harvard faculty member James was instrumental in establishing scientific psychology as an academic discipline in the US. Known as the Father of American Psychology, James founded the country's first psychology laboratory in 1875; taught its first physiological psychology course in 1875/1876; supervised its first PhD granted in psychology in 1878; and published its first important treatise on psychology, *The Principles of Psychology*, in 1890. The latter book is noteworthy in that it was the first psychology written from an overtly non-metaphysical ('positivist') and Darwinian perspective. The *Principles* sketched a definite evolved role for consciousness in directing action, and it embedded this conception in a coherent moral psychology. It earned James instant acclaim and was the key psychological text for a generation.

In the 1890s James seemed to depart from psychology, wrapping up his contributions to the field with two revisions of *The Principles of Psychology*. The first was *Psychology: Briefer Course* (1892), a condensation of the longer text affectionately known as 'Jimmy'. The second was *Talks to Teachers on Psychology* (1899), a pedagogical application of James's ideas traditionally bound together with a series of James's lectures at women's colleges titled *Talks to Students on Some of Life's Ideals*. During this time James distanced himself from experimental work by relinquishing control of his laboratory to his colleague Hugo Münsterberg. He then began showcasing his more recognizably philosophical writings, as in his 1897 volume *The Will to Believe*. The latter book's infamous title piece argues that it is sometimes permissible to hold a belief in the absence of coercive evidence, so long as this belief has momentous and irreversible practical consequences for the believer. James's key assumption is that obtaining truth and avoiding error are both legitimate demands, where neither absolutely trumps the other. Belief for James is a *bet*, where an overly risk-averse attitude is as foolish as an overly risky one.

Given this apparent shift in James's interests, psychologists tend to stop paying attention after the *Principles* (and its revisions), whereas philosophers usually start paying attention around 1897. In fact, however, James spent the 1890s pursuing his intertwined interests in abnormal psychology, psychical phenomena and unconscious dynamics. These interests are relevant to philosophers and psychologists alike. Like numerous writers of his time, James held a version of psychodynamic theory or 'depth psychology'.[24] James's psychodynamic interests culminated in his 1896 Lowell Lectures 'On Exceptional Mental States', which he did not publish and which have gone largely ignored.[25] These lectures are noteworthy, however, in that they show James adding a growth-oriented dimension to his moral psychology. Concepts of conversion and growth are also prominent in *The Varieties of Religious Experience* (1902), a wide-ranging study of individual religious experience. The *Varieties* inaugurated the modern psychology of religion and contains James's first discussion of a 'moral equivalent of war', his proposal for the systematic use of martial energies for constructive ends.[26] The latter concept was a direct influence on both Depression-era public works projects and the Peace Corps in the US.

James was a born popularizer who was more at home on the lecture circuit than in technical expositions of his ideas. This made him a great public intellectual after the manner of Emerson, but it also encouraged a folksy presentation of concepts that is easy to misrepresent. This folksiness is at its height in his discussions of the revered philosophical topic of truth. James's pronouncements on truth in *Pragmatism* (1907) and *The Meaning of Truth* (1910) became emblematic of the school of pragmatism, the other classic figures of which are John Dewey and (more ambivalently) Charles Sanders Peirce. James seems positively to enjoy bringing truth down into the dirt. Thus he argues during the second half of his career that truth is just 'what works'. Stylistic considerations aside, it is fair to say that James held that a belief is verified by its practical consequences within experience. This process of verification is literally the act of its becoming true. Truth is thus not a static relation but a process in time. What exactly constitutes truth is left open-ended, however, as there are indefinitely many ways for beliefs to lead from one part of experience to another. Truth is cashed out in terms of practical effects, but the latter are indeterminate and evolving. James thus characterizes his pragmatism as an embrace of 'the open air and possibilities of nature, as against dogma, artificiality, and the pretence of finality in truth'.[27]

The final major piece of James's philosophy is his metaphysics of radical empiricism. First announced in the 1897 preface to *The Will to Believe*, radical empiricism posits a primordial level of reality called 'pure experience'. The term 'experience' here is not meant to convey that reality is subjective. On the contrary, both subject and object are derivative functions of pure experience. In general, relations for James are as ontologically primordial as what they relate. This applies to any particular things or concepts that are related, as well as to the very subject-object relation itself. Key here is that pure experience is self-sufficient and does not require some metaphysical glue – a transcendental ego or an idealist Absolute – to hold it together. Such a glue would only be necessary if disconnection were a given and connection a secondary accomplishment. However, pure experience for James is a dynamically continuous tissue of relations in which both connection and disconnection are implicit. (Pure experience is not the Kantian manifold of sensation that requires synthesis precisely because it lacks any connectedness.) Radical empiricism is the subject of James's only attempt at a systematic technical exposition of his thought, an abortive book project titled 'The Many and the One' that he worked on in 1904/1905.[28] Radical empiricist themes also pervade *A Pluralistic Universe* (1909), as well as James's posthumously published books *Some Problems in Philosophy* (1911) and *Essays in Radical Empiricism* (1912). These final works warn strongly against 'vicious abstractionism', a fallacy where the dynamic flux of experience is reduced to the skeletal abstractions we glean from it.[29] As a self-styled 'pluralist', James denies that reality is fully knowable from any one perspective or with any one concept or system.

It was in this rarefied metaphysical atmosphere that James finished his career and life. He died of heart failure in 1910 after achieving international fame in multiple fields of study.

Centre and structure

The present study excavates a coherent worldview from this rich body of work. This excavation reveals not only a *centre* to James's vision but also a *structure*.

The individual: the centre of James's vision

The centre of James's vision is his analysis of the individual. James cares most about providing a naturalistic account of the role of individuals in their own development and thus in the development of the societies and other systems that embed them. Oddly enough, this means that *the centre of James's vision is precisely his concern with centres of vision*. A 'centre of vision' for James is just the relatively stable cluster of inter-associated cognitive-affective habits that constitutes one's character or self. The self is in this way literally embodied in the nervous system, where it provides a background of meaning and resistance for ideas and actions. Because moral action for James means the wilful mediation of this structure, James's ethics is one of *self-transformation*.

This centring of the self-transforming individual makes James out to be an essentially moral or ethical thinker. This might appear odd. It is tempting to conclude that James had little to say about ethics. After all, James gave just one course on ethical theory in his career, 'Philosophy 4: Ethics – Recent Contributions to Theistic Ethics' (1888/1889).[30] Shortly thereafter, he echoed the main ideas of the course in his only essay on ethical theory, 'The Moral Philosopher and the Moral Life' (1891). The latter essay seems to present a modified version of utilitarianism. This is hardly revolutionary and seems disconnected from James's psychology or from the *ethics of belief* proffered in 'The Will to Believe'. Given such scattershot contributions, 'The Moral Life' would not seem to be one of James's chief concerns.

This impression is utterly wrong, however. James's philosophy is ethical to its marrow, but it is not an act-based ethics like Kantianism or utilitarianism. In defending an individualistic reading of James's ethics, it is worth getting clear on the meaning of 'individual' as it operates here.

First, numerous interesting questions may be raised about biological individuality in general. How has individuality evolved over time?[31] Should individuals be defined as whatever is selected in natural selection – which may be any number of entities in the biological hierarchy – or in more organismic, physiological terms?[32] Such questions are relevant here, but they are viewed through a frankly anthropocentric lens. Special emphasis is placed on a particular kind of organismic individuality, which is that of the mature acculturated *Homo sapiens*. How do humans develop themselves, given their particular capacities for culture and self-reflexive intelligence? What is the moral relevance of such capacities? This anthropocentric emphasis is not meant to imply that humans are absolutely the most interesting creatures, or that their capacities are discontinuous with those of other beings. Nor does it imply that 'human nature' expresses a static essence or soul, or that human nature can be defined independently of the myriad relations into which humans enter (or out of which they resolve). The point, merely, is to address our *concrete extant interest* in examining the relationship between scientific theory and our own self-image. Science purports to tell us what kind of beings we are, and we have a stake in how it does this.

Second, the emphasis on individuals is not meant to imply that individuals are basic or atomistic units that can be understood in abstraction from the systems that embed them. On the contrary, individuals are interesting precisely because they are complex, functionally integrated wholes that are constituted as such in relation to a lived world. Individuals are thus constituted at the intersection of the twin dialectics of part–whole and self–world.[33] This is the dynamically provisional way in which they are real. There may be some ontologically inflated sense of individuality – defined by relation to an essence, soul or transcendental ego, for instance – in contrast to which this dialectical conception looks anaemic. This does not make individuality an empty fiction, however. On the contrary, it raises good questions about how individuality operates concretely. Rather than artificially inflating the individual, let it be deconstructed so that the interesting work of reconstruction can begin.

Third, this book does not prescribe a narrowly selfish worldview. It is a knee-jerk reaction to think that anyone proffering an individual-centred ethics based on Darwinian ideas must be a *social Darwinist*. This is false, and William James is a counterexample. James is not Ayn Rand. Nor is he Herbert Spencer, whom he rebuked precisely for violating what he took to be the real lesson of evolutionism in ethics: No values are given a priori, because values, like organic functions, are indeterminate and change with time. Evolutionism no more entails egocentrism or laissez-faire capitalism than it entails other forms of ethical or social organization. Since values are not pre-given, we are left with the difficult work of sorting them out in practice. This sorting occurs at the individual level in the form of self-transformation, or the wilful mediation of embodied habits. It simultaneously occurs at the social level, which can be imagined as a set of habits that are mediated by the self-transformative activities of individuals. The social and individual levels may reinforce or weaken habits in each other. In line with the dialectical view suggested above, influence is neither wholly top-down nor bottom-up.

Fourth, there is a long and venerable tradition of writing about self-cultivation and self-care, stretching back to the ancient stoics. Ideally, this book may be viewed as a modern extension of this tradition, rather than as a contribution to the literature on 'self-help' in the current sense of the term. This book is not about maximizing your productivity in a vulgar economic sense or helping you to lose weight. That is, it is not about engraining your existing neuroses so that you may focus on them more narrowly. Frenetic activity without critical self-awareness is not self-helpful behaviour. Nor is it other-helpful. A philosophy of self-transformation should cultivate a tension between existing norms and posited ideals, including rethinking the very structures that generate one's extant fears and desires.

Finally, the intention is surely not to promote obsessive self-involvement or narcissism, which is actually a fundamental kind of existential alienation. Social animals do not flourish in an echo chamber. In the words of John McDermott, 'we must maintain that it is a "world" which we wish to create. We seek more than a dazzling array of self-preening evocations of the human psyche'.[34]

Selectionism: the structure of James's vision

If the centre of James's vision is the individual, then the structure of his vision is *selectionism*. Selectionism is a logic that James gleans from Darwin's theory of natural selection and then applies in various domains. Simply put, this logic explains change in a system in terms of selection upon variation by an environment. James in this way posits the selection of individuals by society; of sensory input by various levels of cognitive processing; of ideas by the will; of behaviours by the environment; and of truths by individuals and societies. The point here is not the well-worn one that James emphasizes *selective attention* in his psychology – although this is part of it. It is also that selective attention, like other instances of selection in James's work, fits a general pattern that has specific logical features.[35]

Why this enthusiasm for selection? The answer is that this logic limits the role of the environment. This might sound strange, if we are used to thinking of natural selection as a process of mechanical winnowing by external conditions. Selectionism has a particular logical feature that James used to great effect, however. This is the 'random' or 'accidental' nature of variation – which may be better termed 'non-directedness' due to the unnecessary connotations of the former terms.[36] Variation is non-directed if it is not directly produced or elicited by environmental demands in a systematically adaptive fashion. Non-directed variation is offered to an environment that did not order it up. James develops his selectionism in response to the externalist or 'outside-in' character of Herbert Spencer's Lamarckism. Evolution for Spencer is a cumulative process of organic adjustment to an autonomous environment, where the adaptive gains of each generation are passed on to the next through the inheritance of acquired characteristics. James employs selectionism to drive a logical wedge between the environment and the sources of variation – a crack through which inscrutable novelty may enter.[37] This is key: Selection is a *deflationary* concept for James, as it signifies a certain lack of power. To select is *merely* to select, not to elicit or produce.

Nevertheless, James does recognize that the environment may exert indirect influence over variation. This is because the environment's selective activities alter the conditions for future variation by biasing the system in favour of particular possibilities. To select a piece of morphology in organic evolution is to constrain the possibility space for future variation. Not every new trait will be developmentally compatible with the newly defined evolutionary space. This indirect biasing of variation is characteristic of the environment in Darwinian biology, as well as of the will in James's ethics. The will on James's view crowns the sensorimotor system, where it selects upon ideas that represent competing possible actions. The will thus comprises a selective environment *within the individual*, analogous in function to the Darwinian environment. Like any selective environment, the will is incapable of producing the variation upon which it selects. However, it does bias future variation in a twofold sense. First, it alters the external social-natural world through its selection of overt actions; and second, any volition – including a decision to dampen an impulse and thus to preclude an overt action – alters the internal habitual environment through the differential weighting of nervous channels. James thus emphasizes what biologists today call 'niche construction', or organisms' alteration of their own environments, with the twist that one half of a moral agent's twofold environment is the internal structure provided by its own nervous system. By breaking or entrenching habitual pathways, one engages in literal self-transformation.

This idea that various systems are similar to natural selection has a long history and has generated considerable interest in the philosophy of science.[38] This idea goes back at least to Darwin himself, who suggested in the *Origin* that languages have evolved in a selectionist manner. Selectionist explanations may be contrasted to *instructionist* ones, which rely on variation that is directed in just the sense that selectionist variation is not. For instance, Lamarckism is an instructionist theory because it posits variations that are prompted directly by environmental demands. The triumph of Darwinian selectionism over Lamarckian instructionism is a key feature of the narrative of modern biology. With all due deference to epigenetic theory, we now explain evolution in terms of the shifting composition of populations of differently endowed individuals, not the cumulative adaptations acquired by individuals during their lives. The giraffe's long neck is due to the outcompeting of shorter giraffes by taller giraffes, not the parallel efforts of all giraffes to get taller.[39]

A less shop-worn example of a selectionist triumph comes from the study of the vertebrate immune system. The immune system destroys invading bodies (pathogens) by binding specialized cells (antibodies) to specific sites on these bodies (antigens). Before the middle of the twentieth century, researchers believed that the immune system learns to target specific pathogens by creating templates of their antigens and using these templates to create the appropriate antibodies. This would be an instructionist process where variation is induced through the direct transfer of structure onto the system from an external source. According to the newer 'clonal selection theory', however, the immune system actually operates according to multiple selectionist processes: First, a rich stock of antibody-variants is produced early in development and then winnowed down so as to eliminate those that would attack friendly somatic cells; and second, a process called 'affinity maturation' favours the reproduction of extant antibody types that react successfully with detected antigens. Thus, just as natural selection explains adaptation in terms of the differential survival and reproduction of organisms with different traits, immunology now explains ontogenetic adaptation in terms of the differential survival and reproduction of antibodies of different types.[40] Other twentieth-century selectionist theories include operant behaviourism in psychology, which explains individual learning in terms of selectionist trial and error; Donald Campbell's evolutionary epistemology, which is a selectionist theory of knowledge-growth; the theory of memes, which are supposed to be gene-like units of cultural selection; and Gerald Edelman's neural Darwinism, which accounts for neurophysiological development by invoking the selective reinforcement and degeneration of neural pathways.[41]

This flowering of selectionist explanation does not imply that selectionism explains everything in complex systems or that it always represents an improvement over instructionism.[42] Selectionism and instructionism are two tools in a toolkit. Nor are they the only tools. For instance, there are also explanations given in terms of internal structures, processes or constraints.[43] Whereas selectionism and instructionism necessarily appeal to external factors (though in different ways), such explanations invoke features that are internal to the system of study. As mentioned above, evolutionary possibilities are constrained by the entrenched structure passed down in a lineage. Horses, for instance, will not be evolving wings. This is not because wings would confer no selective advantage, but because the horse body plan offers no viable way of developing them. This is an internal constraint. There are also highly general physico-chemical constraints defining the possibility space of variation at the outset, as argued by Stuart Kauffman.[44] Insides can explain other insides while reducing the explanatory power of outsides.

What the above array of selectionist explanations does show is that this model may be applied at various levels and in various domains. One debate around selectionism is the 'units of selection' controversy among philosophers of biology. This debate concerns which levels of the biological hierarchy – genes, organisms, or higher-level groups – are subject to evolution by natural selection.[45] Gene-centric reductionists, for example, countenance only the level of the (evolutionarily defined) gene, whereas others may take a more hierarchical view that posits selection at multiple levels of analysis. This is a debate about evolutionary biology. Selectionism may also be applied to different aspects of organic or inorganic systems, however.[46] Developmental biologists, for example, may posit selectionist processes that structure ontogeny, and they may be interested in such processes regardless of any implications they have for phylogeny. Sometimes ontogeny is interesting for ontogeny's sake.[47] One can even imagine a hierarchical selectionist analysis – such as James's construal of the sensorimotor system or 'reflex arc' – that aims principally to say something about moral psychology or agency.

Selectionism is thus not inherently about literal Darwinism or even biology. What selectionism is about depends upon one's purpose for making the analysis in the first place. It is merely a contingent fact of history that Darwin was an important innovator of selectionism. Indeed, selectionist theories need not be conceived as *analogies* to natural selection at all. Instead, natural selection and other selectionist theories can all be measured against an idealized general model that they instantiate.[48] It is even consistent to reject natural selection while accepting other selectionist theories. For these reasons, 'selectionism' is here preferred to other terms such as 'Universal Darwinism' and 'Generalized Darwinism' that include honorific references to Darwin.[49]

Finally, social scientists and humanists have legitimate concerns about encroachments of biological ideas upon their domains (as described in Chapter 2). Such movements as sociobiology and evolutionary psychology in particular raise the spectre of a simplistic reductionism that threatens to swallow other disciplines whole. One need not swear off all interdisciplinary exchange, however. Drawing upon evolutionary theory only risks reductionism if one is drawing up on reductionistic views. Evolutionism in the broadest sense entails neither reductionism nor mechanism at the expense of other explanatory or interpretive modes, even if it requires some historical perspective to be reminded that this is so. Indeed, the present study uses James's evolutionism to complexify and problematize rather than to simplify or reduce.

The literatures

This book lies at the intersection of multiple rich literatures, including James scholarship and various works connecting American pragmatism to the history and philosophy of science.

James scholarship

This is not the first work to attempt to locate the centre of James's vision. For instance, Ellen Suckiel argues in *The Pragmatic Philosophy of William James* (1982) that James's philosophy rests on his twin ideas that humans are inherently purposive and that reality is that which is capable of being experienced; and Charlene Seigfried claims in *William James's Radical Reconstruction of Philosophy* (1990) that James is concerned with 'establishing a secure foundation in experience which would overcome both the nihilistic paralysis of action and the skeptical dissolution of certain knowledge brought on by the challenge of scientific positivism'.[50] This book is like Suckiel's in that it foregrounds the purposive individual in James's work, and it is like Seigfried's in that it reads James's concerns about science as stemming from his concerns about individual agency.

Nevertheless, the reading of James as a philosopher of self-transformation is a heterodox one that is not well represented in a prior generation of scholars. The centrality of self-transformation in James's work is now being recognized by a handful of projects – including the present one – that are arising relatively independently in various countries.[51] When James's philosophy is viewed in terms of self-transformation, James's essay 'The Moral Philosopher and the Moral Life' appears not as a tweak to utilitarianism but as a radical critique of the very possibility of ethical theory;[52] and both the latter essay and 'The Will to Believe' are profitably read as extensions of the moral psychology contained in *The Principles of Psychology* and *Talks to Teachers on Psychology*.

This reading also provides one way of responding to the idea that James's thought is simply incoherent – an issue forced by Richard Gale's provocative *The Divided Self of William James* (1999). Perhaps the centre does not hold. According to Gale, James's writings contain a tension between 'Promethean pragmatism' and 'anti-Promethean mysticism'. On the former view, the individual is active, purposive and wilful. On the latter view, the individual is passive, receptive and capable of mystical union with reality. Whereas the former understands concepts instrumentally and relativizes ontological claims to purposes, the latter identifies truth with the direct perception of present reality. Gale construes this tension as fundamental and irrevocable. Such an interpretation may have value in reminding us that we cannot ontologically have everything we want. However, it can also challenge interpreters to show how different intellectual tendencies are in fact compatible or even dialectically intertwined. The latter perspective is adopted here.

Two relatively recent book-length works on James that have affinities to the present study are James Pawelski's *The Dynamic Individualism of William*

James (2007) and Sergio Franzese's *The Ethics of Energy* (2008).[53] These interpretations are refreshing. It is tempting to periodize James artificially, as if he skipped from psychology to 'The Will to Believe' to occupying himself with religion, truth and metaphysics. However, both Pawelski and Franzese resist this temptation by placing a physiologically grounded analysis of the individual at the centre of James's philosophy. The present work differs from Pawelski and Franzese in that it makes more of James's evolutionism than they do, and in that it gives pride of place to generalized selectionism in James's thought (as opposed to the reflex action model that Pawelski treats as more primary). However, it does agree with them in reading James through the optic of what might be termed *philosophical anthropology*, or the non-reductionistic study of the human being as an indeterminate purposive animal.[54] This viewpoint is tonic, not only for James scholars, but for philosophy more generally. Philosopher of biology Lenny Moss, for instance, has argued for a return to the twentieth-century German tradition of *philosophische Anthropologie*, as represented by thinkers like Max Scheler, Helmuth Pleßner, Kurt Goldstein and Arnold Gehlen. Such a viewpoint avoids a 'vulgar Darwinism' that, like vulgar Marxism, denies agency to individuals in the systems that embed them.[55]

This anthropological optic makes James's contemporary Nietzsche a useful point of comparison. Both James and Nietzsche are radical philosophical anthropologists who critique the tradition from the perspective of a new philosophy of the self-fashioning organism. Indeed, both forge such a philosophy in direct opposition to nineteenth-century evolutionary and physiological logics that portray that individual as passive relative to its circumstances. Where James and Nietzsche come apart, however, is in their normative visions for what constitutes an ideal character or society: James posits a utopia of widely shared flourishing, while Nietzsche invokes a form of exemplary individuality that stands above society and apart from it. If we take seriously the idea that no ultimate values are pre-given in a post-Darwinian worldview – or in the wake of what Nietzsche calls the 'death of God' – then adjudicating among such ideals becomes a difficult but urgent matter.

James and Rorty: truth de-centred

Reading James as a philosopher of self-transformation reorients other areas of his thought. For example, it sheds different light on his theory of truth. Although this is one of the most examined aspects of James' philosophy, it does not represent a strong alternative centre of James's vision in comparison to his ethics of self-transformation. This is because James's theory of truth is not a competitor to his practical philosophy but rather one of its facets. A brief comparison with neo-pragmatist Richard Rorty may help to motivate this approach, given Rorty's position as a well-known philosophical provocateur whose writings feature a similar (but differently motivated) decentring of truth.

Rorty launched a critique of correspondence theories of truth in *Philosophy and the Mirror of Nature* (1979). This book deconstructs a central metaphor of the philosophical tradition: namely, that mind is a mirror for representing the world. This metaphor implies that knowledge consists of a collection of representations, where philosophy is the practice of polishing and maintaining the mirror that reflects them. According to Rorty, this metaphor introduces a raft of pseudo-problems, including the very idea that knowledge is a special kind of subject matter requiring its own theory. Rorty arrives at this position through a sweeping critique of modern philosophy. For instance, he rejects the empiricist view ascribed to John Locke, on which knowledge means having one's senses impacted in a special manner by external objects. The problem with such views is that they provide a causal story instead of a justificatory one. To use the language that Rorty adopts from Wilfrid Sellars, justification occurs in the conceptual 'space of reasons', which is incommensurate with the 'space of causes' that we invoke in explaining our physical transactions with the world. If justifications are necessarily conceptual in form, then only an encultured conversation partner – not a brute causal nexus – can be said to justify. Rorty concludes that philosophy-as-epistemology should be supplanted by a cultural hermeneutics that does not ground justification in anything non-human or extra-linguistic. In this regard, his project has affinities to the poststructuralist movement that was also gaining steam in the 1970s and 1980s.

Rorty's book brought the pragmatist critique of correspondence theories back to the attention of the philosophical community, where it had been largely ignored since the death of John Dewey in the middle of the twentieth century. It was James, however, who inaugurated this critique. A full century before *Philosophy and the Mirror of Nature*, James published 'Remarks on Spencer's Definition of Mind as Correspondence' (1878), a critique of Herbert Spencer's view that the function of cognition is the passive formation of psychological associations that mirror relations in the external world. As against this view, James contends that the individual actively biases its own sensory and perceptual experience while also generating novel ideas and actions that are not merely the resultant of environmental pressures. James's point in this essay is not only that Spencer's account is too linear and thus descriptively incorrect. It is also that Spencer is smuggling an unwarranted epistemic norm into his psychology: To think rightly is to mirror the environment to which one is tethered. In contrast, James argues that no criterion for right thinking can be posited a priori. Rather than being set by a foundational truth or the evolutionary past, the function of thinking depends upon the ongoing activities of the individuals in whose lives cognition is interwoven with practice. This could be described as a 'correspondence theory', only if 'correspondence' means not passive synchronic mirroring but active diachronic *coordination* with a lived world.

Although both James and Rorty reject the idea of truth as a static relation between mind and an authoritative type of object, they arrive at these views by different routes. As Rorty stresses in his later writings, he is essentially an anti-authoritarian thinker who believes that correspondence theories are, like theism, a way for humans to debase themselves before a supposed non-human power. Reared in analytic epistemology, Rorty concludes that this field should be razed in order to sow the seeds of an anti-foundationalist democratic pluralism. In contrast, James was a trained psychologist who disclaimed representationalist empiricism in favour of a post-Darwinian belief in the open-endedness of function and value. In short, Rorty's duality is *authoritarianism versus democracy*, and his heresy of analytic philosophy supports the latter; James's duality is *externalism versus individualism*, and his heresy of empiricist psychology supports the latter.

History and philosophy of science

In examining James's heresy of psychology, the present study builds upon multiple extant works that situate pragmatism in the history of science. The classic book on evolution and pragmatism is Philip Wiener's *Evolution and the Founders of Pragmatism* (1949). This book is by now quite dated, however, and it lacks the conceptual apparatus of today's philosophy of biology. More current is Robert Richards's impressive *Darwin and the Emergence of Evolutionary Theories of Mind and Behavior* (1987), a detailed examination of post-Darwinian psychology that examines James among others. The present study's framing of a Pragmatic Image of Darwinism in particular squares with Richards's emphasis on the importance of behavioural and moral theories in nineteenth-century evolutionism. This book is also indebted to Peter Godfrey-Smith's *Complexity and the Function of Mind in Nature* (1996), from which it adopts the framework of internalism, externalism and constructionism. In addition, a relatively obscure essay by information theorist Jonathan Schull helped to inspire the idea of a hierarchical selectionism in James's work.[56] Finally, Paul Croce's *Science and Religion in the Era of William James* (1995) is useful in that it makes James the protagonist in the story of the rise of uncertainty in nineteenth-century science.

A growing body of more recent work approaches the classical pragmatists with the sharpened tools of today's history and philosophy of science: Both Trevor Pearce and Michael Brady present book-length studies of the pragmatists within the broader history of American evolutionism, which are welcome updates to Wiener's seminal work;[57] Alexander Klein investigates James's empiricism in relation to his psychology and philosophy of science;[58] and Rasmus Grønfeldt Winther examines James's and Dewey's respective warnings about the limits of scientific abstraction.[59] Such studies usefully dissect old thinkers along new lines, exposing new functional interrelations within their thought.[60]

In so dissecting James, the present study aims to secure him a more prominent seat at the Darwinian table. That is, it challenges an assumption that is enshrined in the title of Jerome Popp's book *Evolution's First Philosopher: John Dewey and the Continuity of Nature* (2007). Dewey's Darwinism is indeed a rich area of study, especially as it combines evolutionary naturalism with Hegelian insights about the dialectical relationship between parts and wholes, or between organism and environment.[61] Dewey is not evolution's first philosopher, however. At a minimum, to make this claim ignores two important books that happen to share a name: *The Principles of Psychology*. The 1855 first edition of

Herbert Spencer's *Principles of Psychology* predates Dewey's birth. If anyone was evolution's first philosopher, it was Spencer. Spencer formulated an entire evolutionary system comprising an evolutionary cosmology, biology, psychology and sociology. He was an evolutionist before Darwin published the *Origin* and popularized the very word 'evolution'. In doing so, he was engaged in ongoing debates with leading scientists including Darwin, Huxley and August Weismann. He was also the evolutionist of choice for the general public, publishing countless popular books in his series of 'Synthetic Philosophy'. These works influenced American thinkers including James and his companion, the evolutionary philosopher John Fiske. Even if by arbitrary fiat 'evolutionism' is defined as 'Darwinism' in order to rule out the Lamarckian Spencer, Dewey was not, as Popp claims, the first to find in Darwinism a naturalistic basis for meaning and value. Indeed, the present study can be viewed as an extended examination of James's efforts in doing just this. James's 1890 *Principles of Psychology* – instrumental in turning Dewey toward evolutionary naturalism in the first place – is a landmark text in developing such an account.

Popp's book is also limiting in that it casts Dewey in the mould of reductionists Daniel Dennett and Richard Dawkins. In doing so, it ignores contemporary schools of non-reductionist evolutionism such as dialectical biology and developmental systems theory. The latter schools are much closer to a Pragmatic Image of Darwinism – or to James or Dewey – than are Dawkins or Dennett. They are pluralistic outlooks that buck the neo-Darwinian trend by emphasizing nonlinear causation, non-genetic modes of heredity and an open-ended diversity of explanatory factors.[62] In doing so, they open a role for the organism as such, among a host of other factors, in accounts of both ontogeny and phylogeny. They are therefore important points of reference in the present study. Dennett and Dawkins also play a role, but primarily as foils.

Plan of the book

This book unearths a strong backbone running through James's psychological and philosophical writings. In doing so, it comprises a wide-ranging but historically grounded study of the reciprocal influence of scientific theorizing and humanity's ethical self-conception.

Chapter 1 investigates James's Darwinian psychology. James was the first important Darwinian psychologist, but his view differs importantly from today's neo-Darwinism. Darwin's theory of natural selection rested on uncertain probabilistic evidence and could only be confirmed in a holistic manner. Nevertheless, James embraced this theory and generalized its logic of non-directed ('random') variation in order to argue that individuals are replete with internally generated idiosyncrasies that influence a range of systems. The twin lessons of Darwinism for James – in contrast to neo-Darwinism – are that knowledge is inherently uncertain and that individuals are a real locus of agency in the world.

Chapter 2 examines James's social evolutionism. Evolutionary theory has arisen through a complex dialectic among the physical, social and life sciences.

Such interdisciplinary exchange has been the source of real insights, as in Darwin's generalization of concepts from classical economic theory. It has also generated confused and dangerous positions, however, such as social Darwinism and reductionistic sociobiology. William James's social evolutionism is a useful corrective to such views. The point of James's social evolutionism is not to prescribe struggle or to absorb culture into biology, but rather to insist upon the complex, hierarchical and nonlinear quality of the systems in which individuals are embedded.

James's ethics of self-transformation takes centre stage in Chapter 3. Physiology need not reduce or eliminate other approaches to mind or behaviour. Instead, it can be one facet of a broader study of the human being as a concrete purposive organism. James put forth just such a view in introducing the study of physiological psychology to the US in the 1870s. According to James, the human being comprises a hierarchical sensorimotor system that is structured by plastic habits and mediated by a selective will. This structure is the basis of James's ethics, on which moral action means resisting entrenched habitual structures for ideal ends.

Chapter 4 brings James's evolutionary worldview into conversation with that of his German contemporary Nietzsche. James and Nietzsche each built radical philosophies based upon an analysis of the individual as a self-fashioning organism. Both agree that a meaningful life must employ energetic and form-giving powers, and both analyse the individual as a set of hierarchically ordered subselves. Both also utilize Darwinian concepts in reconstructing individual agency, although Nietzsche's relationship with Darwinism is more antagonistic than James's. In the end, James's goal is to enact a utopia that is hospitable to humanity's extant social and metaphysical needs, whereas Nietzsche seeks to cultivate an exemplary post-humanity that digests those needs through ascetic self-overcoming. These are starkly different options for the reconstruction of humanity in the wake of Darwinism's deconstruction of teleology and Nietzsche's proclaimed death of God.

Chapter 5 broadens the historical and conceptual scope of the project to include James's mature concepts of pragmatism, pluralism and radical empiricism. If nineteenth-century idealism construes reality as an *absolute organism*, then James construes reality as a *finite relative organism* – that is, the only kind of organism that has ever been observed. According to James's pragmatism and radical empiricism, individuals are organisms fringed by an inchoate 'more' that is never quite grasped, and we reside in a similarly finite, fringed and growing world. Truth itself is a growing aspect of a developing reality. At his most speculative, James argues that the world is continuously infused with novelty that is mediated by a hierarchy of minds 'summing up' to a finite God. Here James's ethics of self-transformation appears in its broadest logical and metaphysical terms.

The book concludes by taking stock of some schools of thought that are allies to James's Pragmatic Image of Darwinism: dialectical biology, developmental systems theory, autopoesis theory and humanistic psychology. Each of these schools of thought helps us to avoid what James calls 'vicious abstractionism', or the method of treating a concept or theory as exhaustive of the reality that it purports to explain. Dialectical thinking in particular is offered as an antidote to vicious abstractionism. James was ambivalent toward dialectics in its Hegelian guise, but its logic captures something of the dynamic and nonlinear quality that he was attempting to get at in his selectionism. Individuals for James are centres of inchoate possibility within a genuinely unfinished universe constructed through multiple interwoven dialectics.

Notes

1 Darwin 1998/1859, 88.
2 Sober (2009) argues that Darwin wrote the *Origin* 'backwards' by placing his argument for the importance of natural selection up front, before his evidence for the branching pattern of common descent. This is because the pattern provides the background against which the process is intelligible and explanatory.
3 Strictly speaking, Darwin divides artificial selection into methodological selection and unconscious selection. Only the former is intentional. The latter occurs when breeders favourably treat individuals that have desired traits, but with 'no wish or expectation of permanently altering the breed' (Darwin 1998/1859, 54).
4 Dewey 1910.
5 Mayr 1976.
6 For instance, the common *cladistic view* defines species purely in terms of their position in the phylogenetic tree of life. In contrast, the *homeostatic property cluster* view defends a conception of species as natural kinds. Here species are defined in terms of typical traits that are clustered due to causal mechanisms, where no given trait is necessary for species membership (Boyd 1999). Finally, many philosophers now argue that species are not natural kinds but spatio-temporal *individuals* (Ghiselin 1974; Hull 1978). On this view, phylogeny is like ontogeny in that each consists in the development of a real concrete individual from birth (speciation) to death (extinction).
7 Krüger *et al.* 1987.
8 Ghiselin 1969; Ayala 2009; Sober 2009.
9 Dennett 1995.
10 This framing is adapted from Richards's (1987) concept of the Received View of Darwinism.
11 The school of 'new mechanism' embraces this position, explaining as many systems as possible in mechanistic terms. Natural selection is a potential stumbling block for this view, however. Can a glacially paced stochastic process that lacks tightly integrated physical components count as a mechanism? Is natural selection best described as a *driver* of evolution (mechanistically or not), or as a statistical epiphenomenon? For a critique of natural selection as a mechanism, see Skipper and Millstein (2005). For a defence, see Barros (2008). For an overview of new mechanism, see Levy (2013).
12 Freud believed that even his most poetic theories would eventually be reduced to neurophysiological terms. See Freud's (1966/1895) manuscript 'Project for a Scientific Psychology'.
13 Taylor 2001.
14 The distinction between internalism, externalism and constructionism is adopted from Godfrey-Smith (1996).
15 This sketch is cursory. James's early authoritative biographer is Perry (1935). See also Richardson (2006). Broader intellectual histories that give James a prominent role include Croce (1995) and Menand (2002). For a history of the James family extending back to eighteenth-century agrarian Ireland, see Lewis (1993).

16 PU, 44.
17 PU, 117.
18 LWJ II, 355. That James himself was adept at such vision-catching is captured by a remark by his friend Peirce: 'His comprehension of men to the very core was most wonderful ... in all my life I found scarce any soul that seemed to comprehend, – naturally, my concepts, but the mainspring of my life, better than he did' (quoted in Perry 1935 I, 541).
19 Foucault 1977.
20 Such early influences do not map easily onto the most commonly taught traditions in academic philosophy departments. This makes it dangerous to pigeonhole William James into familiar categories. This problem is compounded by the fact that James was largely self-taught in philosophy and took seriously fringe thinkers – such as gas-huffing pamphleteer Benjamin Blood – who are today virtually unknown.
21 Perry 1935 I, 228.
22 'Remarks on Spencer's Definition of Mind as Correspondence' (EPH, 7–22).
23 'Great Men and Their Environment' (WB, 163–189).
24 The various depth psychologies were supplanted by the monolith of Freudianism in the twentieth century. See Taylor 2002.
25 Taylor 1984.
26 The proposal is developed in James's 1910 paper 'The Moral Equivalent of War' (ERM, 162–173).
27 P, 31.
28 MEN, 3–53.
29 Abstractionist fallacies in James and Dewey are explored by Winther (2014).
30 MEN, 182–185.
31 Buss 1987.
32 Pradeu 2010.
33 Levins and Lewontin 1985; Varela 1991.
34 McDermott 1976, 44.
35 Schull 1996.
36 Amundson 1989.
37 Crippen (2011) places the non-directedness of mental variation at the centre of James's thought in a manner that is consonant with the present study.
38 Lewontin 1970; Dawkins 1989/1976; Hull 1980; Sober 1984; Edelman 1987; Amundson 1989; Darden and Cain 1989; Cziko 1997; Griesemer 2005; Aldrich et al. 2008; Godfrey-Smith 2009. More historical accounts include Campbell (1974) and Hodgson (2005).
39 Levins and Lewontin 1985, 85–88; Sober 1984, 149.
40 For more on the immunological example, see Pradeu (2010).
41 For a comparison of several of these theories, see Hull et al. (2001).
42 Cziko (1997) adheres to something like this teleological view, where selectionism represents the maturity of a science. Amundson (1989) is more measured about the promise of selectionism.
43 Winther and Oyama 2001.
44 Kauffman 1995.
45 Lewontin 1970; Hull 1980. As Lloyd (2001) argues, this debate can be unpacked into several distinct questions: What is *replicated* in evolution? What *interacts* with the environment so as to cause the differential replication by which evolution is typically measured? What is *benefiting* from this process? What *manifests the adaptations* resulting from this process?
46 Bickhard and Campbell 2003.
47 Amundson 1994.
48 Amundson 1989.
49 For 'Generalized Darwinism', see Aldrich et al. (2008). For 'Universal Darwinism',

see Dawkins (1983) and Nelson (2007). See also 'Darwinian Populations' in Godfrey-Smith (2009).
50 Seigfried 1990, 2.
51 The term 'self-transformation' here is owed to Colin Koopman (personal communication; forthcoming). This term and idea also appears in Uffelmann (2011) and Marchetti (2015a). See also subsequent citations.
52 See Franzese (2008) and James's own statement at the beginning of his essay.
53 Marchetti's book (2015b) is also relevant, but the related article (Marchetti 2015a) is closer to the present study's focus on self-transformation. (The book is about developing a broader conception of philosophical critique.) Slater's book (2009) reads James in terms of flourishing or eudemonia, but more in the context of religion than Darwinism or self-transformation. Other books on James's ethics include Brennan (1961), Roth (1969) and Throntveit (2014).
54 James has been called a philosophical anthropologist as early as the 1960s: Edie (1965) applies this label to James and emphasizes his European influences and resonances; according to DeArmey (1978; 1987), James's philosophical anthropology unifies phenomenological and scientific investigations; finally, Finnish scholar Sami Pihlström (1998; 2007) has formulated a Jamesian culture-oriented philosophical anthropology.
55 Moss 2005.
56 Schull 1996.
57 Brady 2013; Pearce (in preparation).
58 Klein (in preparation).
59 Winther 2014.
60 Further texts on James and Darwinism include Seigfried (1984), Taylor (1990), Bredo (2002), Crippen (2010), Franzese (2009), and Shaw (2010).
61 Pearce 2014.
62 For dialectical biology, see Levins and Lewontin (1985). For developmental systems theory, see Oyama (2000a/1985), Oyama (2000b), and Oyama *et al.* (2003). See also the related niche construction theory of Odling-Smee *et al.* (2003) and Laland *et al.* (2003).

References

Aldrich, H. E., G. M. Hodgson, D. L. Hull, T. Knudsen, J. Mokyr and V. J. Vanberg. 2008. In Defence of Generalized Darwinism. *Journal of Evolutionary Economics* 18, no. 5: 577–596.

Amundson, R. 1989. The Trials and Tribulations of Selectionist Explanations. In *Issues in Evolutionary Epistemology*, eds K. Hahlweg and C. A. Hooker, 413–432. Albany, NY: State University of New York Press.

Amundson, R. 1994. Two Concepts of Constraint: Adaptationism and the Challenge from Developmental Biology. *Philosophy of Science* 6, no. 4: 556–578.

Ayala, F. J. 2009. Darwin and the Scientific Method. *Proceedings of the National Academy of Sciences* 106, Supplement 1: 10,033–10,039.

Barros, D. B. 2008. Natural Selection as a Mechanism. *Philosophy of Science* 75, no. 3: 306–322.

Bickhard, M. H. and D. T. Campbell. 2003. Variations in Variation and Selection: The Ubiquity of the Variation-and-Selective-Retention Ratchet in Emergent Organizational Complexity. *Foundations of Science* 8, no. 3: 215–282.

Boyd, R. 1999. Homeostasis, Species, and Higher Taxa. In *Species: New Interdisciplinary Essays*, ed. R. Wilson, 141–185. Cambridge, MA: MIT Press.

Brady, M. 2013. Evolution and the Transformation of American Philosophy. PhD thesis. Southern Illinois University at Carbondale.

Bredo, E. 2002. The Darwinian Center to the Vision of William James. In *William James and Education*, eds J. Garrison, R. Podeschi and E. Bredo, 1–26. New York: Teachers College Press.

Brennan, B. P. 1961. *The Ethics of William James*. New York: Bookman Associates.

Buss, L. W. 1987. *The Evolution of Individuality*. Princeton, NJ: Princeton University Press.

Campbell, D. T. 1974. Evolutionary Epistemology. In *The Philosophy of Karl Popper*, ed. P. A. Schilpp, 413–463. La Salle, IL: Open Court.

Crippen, M. 2010. William James on Belief: Turning Darwinism against Empiricistic Skepticism. *Transactions of the Charles S. Peirce Society* 46, no. 3: 477–502.

Crippen, M. 2011. William James and His Darwinian Defense of Freewill. In *150 Years of Evolution: Darwin's Impact on Contemporary Thought and Culture*, eds M. R. Wheeler and W. A. Nericcio, 69–89. San Diego, CA: San Diego State University Press.

Croce, P. J. 1995. *Science and Religion in the Era of William James: Volume 1, Eclipse of Certainty, 1820–1880*. Chapel Hill, NC: University of North Carolina Press.

Cziko, G. 1997. *Without Miracles: Universal Selection Theory and the Second Darwinian Revolution*. Cambridge, MA: MIT Press.

Darden, L. and J. A. Cain. 1989. Selection Type Theories. *Philosophy of Science* 56, no. 1: 106–129.

Darwin, C. 1998. *The Origin of Species By Means of Natural Selection, Or The Preservation of Favored Races in the Struggle for Life*. 6th edn. New York: Modern Library. Original edition, 1859.

Dawkins, R. 1983. Universal Darwinism. In *From Molecules to Men*, ed. D. S. Bendall, 403–425. New York: Cambridge University Press.

Dawkins, R. 1989. *The Selfish Gene*. 2nd edn. Oxford: Oxford University Press. Original edition, 1976.

DeArmey, M. H. 1978. The Philosophical Anthropology of William James: Towards a Complete Teleological Analysis of the Nature, Origin and Destiny of Human Beings. PhD thesis. Tulane University, Louisiana.

DeArmey, M. H. 1987. The Anthropological Foundations of William James's Philosophy. In *The Philosophical Psychology of William James*, eds M. H. DeArmey and S. Skousgaard, 17–40. Washington, DC: Center for Advanced Research in Phenomenology & University Press of America.

Dennett, D. C. 1995. *Darwin's Dangerous Idea*. New York: Touchstone.

Dewey, J. 1910. The Influence of Darwinism on Philosophy. In *The Influence of Darwinism on Philosophy: And Other Essays in Contemporary Philosophy*, 1–19. New York: Henry Holt and Company.

Edelman, G. M. 1987. *Neural Darwinism: The Theory of Neuronal Group Selection*. New York: Basic Books.

Edie, J. M. 1965. Notes on the Philosophical Anthropology of William James. In *An Invitation to Phenomenology: Studies in the Philosophy of Experience*, ed. J. M. Edie, 110–132. Chicago, IL: Quadrangle Books.

Foucault, M. 1977. What is an Author? *The Foucault Reader*, ed. P. Rabinow, 101–117. New York: Pantheon Books.

Franzese, S. 2008. *The Ethics of Energy: William James's Moral Philosophy in Focus*. Frankfurt, Germany: Ontos Verlag.

Franzese, S. 2009. *Darwinismo e pragmatismo. E altri studi su William James*. Milan, Italy: Mimesis.

Freud, Sigmund. 1966. Project for a Scientific Psychology. In *The Standard Edition of the Complete Psychological Works of Sigmund Freud, Volume I (1886–1899): Pre-Psycho-Analytic Publications and Unpublished Drafts*, eds J. Strachey *et al.*, 281–391. London: Hogarth Press. Unpublished manuscript, 1895.

Gale, R. M. 1999. *The Divided Self of William James*. Cambridge: Cambridge University Press.

Ghiselin, M. T. 1969. *The Triumph of the Darwinian Method*. Berkeley, CA: University of California Press.

Ghiselin, M. T. 1974. A Radical Solution to the Species Problem. *Systematic Biology* 23, no. 4: 536–544.

Godfrey-Smith, P. 1996. *Complexity and the Function of Mind in Nature*. Cambridge Studies in Philosophy and Biology. Cambridge: Cambridge University Press.

Godfrey-Smith, P. 2009. *Darwinian Populations and Natural Selection*. Oxford: Oxford University Press.

Griesemer, J. R. 2005. The Informational Gene and the Substantial Body: On the Generalization of Evolutionary Theory by Abstraction. In *Idealization XII: Correcting the Model. Idealization and Abstraction in the Sciences*, eds M. R. Jones and N. Cartwright, 59–115. Poznán Studies in the Philosophy of the Sciences and the Humanities. Amsterdam, The Netherlands: Rodopi.

Hodgson, G. M. 2005. Generalizing Darwinism to Social Evolution: Some Early Attempts. *Journal of Economic Issues* 39, no. 4: 899–914.

Hull, D. L. 1978. A Matter of Individuality. *Philosophy of Science* 45, no. 3: 335–360.

Hull, D. L. 1980. Individuality and Selection. *Annual Review of Ecology and Systematics* 11, 311–332.

Hull, D. L., R. E. Langman and S. S. Glenn. 2001. A General Account of Selection: Biology, Immunology, and Behavior. *Behavioral and Brain Sciences* 24, no. 3: 511–528.

James, W. 1975. *The Meaning of Truth*. The Works of William James. Edited by F. Burkhardt, F. Bowers and I. K. Skrupskelis. Cambridge, MA: Harvard University Press. Original edition, 1909.

James, W. 1975. *Pragmatism*. The Works of William James. Edited by F. Burkhardt, F. Bowers and I. K. Skrupskelis. Cambridge, MA: Harvard University Press. Original edition, 1907.

James, W. 1976. *Essays in Radical Empiricism*. The Works of William James. Edited by F. Burkhardt, F. Bowers and I. K. Skrupskelis. Cambridge, MA: Harvard University Press. Original edition, 1912.

James, W. 1977. *A Pluralistic Universe*. The Works of William James. Edited by F. Burkhardt, F. Bowers and I. K. Skrupskelis. Cambridge, MA: Harvard University Press. Original edition, 1908.

James, W. 1978. *Essays in Philosophy*. The Works of William James. Edited by F. Burkhardt, F. Bowers and I. K. Skrupskelis. Cambridge, MA: Harvard University Press.

James, W. 1979. *Some Problems of Philosophy*. The Works of William James. Edited by F. Burkhardt, F. Bowers and I. K. Skrupskelis. Cambridge, MA: Harvard University Press. Original edition, 1911.

James, W. 1979. *The Will to Believe and Other Essays in Popular Philosophy*. The Works of William James. Edited by F. Burkhardt, F. Bowers and I. K. Skrupskelis. Cambridge, MA: Harvard University Press. Original edition, 1897.

James, W. 1981. *The Principles of Psychology*. 2 vols. The Works of William James. Edited by F. Burkhardt, F. Bowers and I. K. Skrupskelis. Cambridge, MA: Harvard University Press. Original edition, 1890.

James, W. 1982. *Essays in Religion and Morality*. The Works of William James. Edited by F. Burkhardt, F. Bowers and I. K. Skrupskelis. Cambridge, MA: Harvard University Press.

James, W. 1983. *Talks to Teachers on Psychology: And to Students on Some of Life's Ideals*. Edited by F. Burkhardt, F. Bowers and I. K. Skrupskelis. Cambridge, MA: Harvard University Press. Original edition, 1899.

James, W. 1985. *Psychology: Briefer Course*. The Works of William James. Edited by F. Burkhardt, F. Bowers and I. K. Skrupskelis. Cambridge, MA: Harvard University Press. Original edition, 1892.

James, W. 1985. *The Varieties of Religious Experience*. The Works of William James. Edited by F. Burkhardt, F. Bowers and I. K. Skrupskelis. Cambridge, MA: Harvard University Press. Original edition, 1902.

James, W. 1987. *Essays, Comments, and Reviews*. The Works of William James. Edited by F. Burkhardt, F. Bowers and I. K. Skrupskelis. Cambridge, MA: Harvard University Press.

James, W. 1988. *Manuscript Essays and Notes*. The Works of William James. Edited by F. Burkhardt, F. Bowers and I. K. Skrupskelis. Cambridge, MA: Harvard University Press.

James, W. and H. James. 1920. *The Letters of William James: Two Volumes Combined*. Boston, MA: Little, Brown, and Co.

Kauffman, S. A. 1995. *At Home in the Universe: The Search for Laws of Self-Organization and Complexity*. New York: Oxford University Press.

Klein, A. In preparation. *The Rise of Empiricism: William James and the Struggle over Psychology*.

Koopman, C. Forthcoming. The Will, the Will to Believe, and William James: An Ethics of Freedom as Self-Transformation. *Journal of the History of Philosophy*.

Krüger, L., L. Daston and M. Heidelberger, eds. 1987. *The Probabilistic Revolution. Volume 1: Ideas in History*. Cambridge, MA: MIT Press.

Laland, K. N., F. J. Odling-Smee and M. W. Feldman. 2003. Niche Construction, Ecological Inheritance, and Cycles of Contingency in Evolution. In *Cycles of Contingency: Developmental Systems and Evolution*, eds S. Oyama, P. E. Griffiths and R. D. Gray, 117–126. Cambridge, MA: MIT Press.

Lamarck, J. B. 1809. *Philosophie zoologique*. 2 vols. Paris: Dentu.

Levins, R. and R. C. Lewontin. 1985. *The Dialectical Biologist*. Cambridge, MA: Harvard University Press.

Levy, A. 2013. Three Kinds of New Mechanism. *Biology & Philosophy* 28, no. 1: 99-114.

Lewis, R. W. B. 1993. *The Jameses: A Family Narrative*. New York: Anchor.

Lewontin, R. C. 1970. The Units of Selection. *Annual Review of Ecology and Systematics* 1: 1–18.

Lloyd, E. A. 2001. Units and Levels of Selection: An Anatomy of the Units of Selection Debates. In *Thinking about Evolution: Historical, Philosophical, and Political Perspectives*, eds R. S. Singh *et al.*, 267–291. Cambridge: Cambridge University Press.

Marchetti, S. 2015a. Unfamiliar Habits: The Ethics and Politics of Self-Transformation. *William James Studies* 11: 102–113.

Marchetti, S. 2015b. *Ethics and Philosophical Critique in William James*. New York: Palgrave Macmillan.

Mayr, E. 1976. Typological versus Population Thinking. *Evolution and the Diversity of Life*, 26–29. Cambridge, MA: Harvard University Press.

McDermott, J. J. 1976. *The Culture of Experience: Philosophical Essays in the American Grain*. New York: New York University Press.

Menand, L. 2002. *The Metaphysical Club: A Story of Ideas in America.* New York: Strauss and Giroux.

Moss, L. 2005. Darwinism, Dualism, and Biological Agency. In *Darwinism and Philosophy*, eds V. Hosle and C. Illies, 349–363. Notre Dame, IN: University of Notre Dame Press.

Nelson, R. R. 2007. Universal Darwinism and Evolutionary Social Science. *Biology and Philosophy* 22, no. 1: 73–94.

Odling-Smee, F. J., K. N. Laland and M. W. Feldman. 2003. *Niche Construction: The Neglected Process in Evolution.* Princeton, NJ: Princeton University Press.

Oyama, S. 2000a. *The Ontogeny of Information: Developmental Systems and Evolution.* 2nd edn. Science and Cultural Theory. Edited by B. H. Smith and R. E. Weintraub. Durham, NC: Duke University Press. Original edition, 1985.

Oyama, S. 2000b. *Evolution's Eye: A Systems View of the Biology-Culture Divide.* Science and Cultural Theory. Edited by B. H. Smith and R. E. Weintraub. Durham, NC: Duke University Press.

Oyama, S., P. E. Griffiths and R. D. Gray. 2003. *Cycles of Contingency: Developmental Systems and Evolution.* Cambridge, MA: MIT Press.

Pawelski, J. O. 2007. *The Dynamic Individualism of William James.* Albany, NY: State University of New York Press.

Pearce, T. 2014. The Dialectical Biologist, circa 1890: John Dewey and the Oxford Hegelians. *Journal of the History of Philosophy* 52, no. 4: 747–777.

Pearce, T. Forthcoming. Pragmatism, Evolution, and Ethics. In *The Cambridge Handbook to Evolutionary Ethics*, eds M. Ruse and R. J. Richards. Cambridge: Cambridge University Press.

Pearce, T. In preparation. *Pragmatism's Evolution: Organism and Environment in American Philosophy.*

Perry, R. B. 1935. *The Thought and Character of William James.* 2 vols. Boston, MA: Little, Brown, and Company.

Pihlström, S. 1998. *Pragmatism and Philosophical Anthropology: Understanding Our Human Life in a Human World.* New York: P. Lang.

Pihlström, S. 2007. Metaphysics with a Human Face: William James and the Prospects of Pragmatist Metaphysics. *William James Studies* 2, no. 1: 1–28.

Popp, J. A. 2007. *Evolution's First Philosopher: John Dewey and the Continuity of Nature.* Albany, NY: State University of New York Press.

Pradeu, T. 2010. What is an Organism? An Immunological Answer. *History and Philosophy of the Life Sciences* 32, nos 2–3: 247–267.

Richards, R. J. 1987. *Darwin and the Emergence of Evolutionary Theories of Mind and Behavior.* Chicago, IL: University of Chicago Press.

Richardson, R. D. 2006. *William James in the Maelstrom of American Modernism: A Biography.* Boston, MA: Mariner Books.

Rorty, R. 1979. *Philosophy and the Mirror of Nature.* Princeton, NJ: Princeton University Press.

Roth, J. K. 1969. *Freedom and the Moral Life: The Ethics of William James.* Philadelphia, PA: Westminster Press.

Schull, J. 1996. William James and the Broader Implications of a Multilevel Selectionism. In *Adaptive Individuals in Evolving Populations: Models and Algorithms*, eds R. K. Belew and M. Mitchell, 243–256. Boston, MA: Addison-Wesley Longman.

Seigfried, C. H. 1984. Extending the Darwinian Model: James's Struggle with Royce and Spencer. *Idealistic Studies* 14, no. 3: 259–272.

Seigfried, C. H. 1990. *William James's Radical Reconstruction of Philosophy*. Albany, NY: State University of New York Press.

Shaw, E. C. 2010. William James on Human Nature and Evolution. PhD thesis, Catholic University of America, Washington, DC.

Skipper, R. A. and R. L. Millstein. 2005. Thinking about Evolutionary Mechanisms: Natural Selection. *Studies in History and Philosophy of Science Part C: Studies in History and Philosophy of Biological and Biomedical Sciences* 36, no. 2: 327–347.

Slater, M. R. 2009. *William James on Ethics and Faith*. New York: Cambridge University Press.

Sober, E. 1984. *The Nature of Selection*. Cambridge, MA: MIT Press.

Sober, E. 2009. Did Darwin Write the Origin Backwards? *Proceedings of the National Academy of Sciences* 106, Supplement 1: 10,048–10,055.

Spencer, H. 1855. *The Principles of Psychology*. 2 vols. London: Longmans.

Suckiel, E. K. 1982. *The Pragmatic Philosophy of William James*. Notre Dame, IN: University of Notre Dame Press.

Taylor, E. 1984. *William James on Exceptional Mental States: The 1896 Lowell lectures*. New York: Scribner.

Taylor, E. 1990. William James on Darwin: An Evolutionary Theory of Consciousness. *Annals of the New York Academy of Sciences* 602, no. 1: 7–34.

Taylor, E. 2001. Positive Psychology and Humanistic Psychology: A Reply to Seligman. *Journal of Humanistic Psychology* 41, no. 1: 13–29.

Taylor, E. 2002. William James and Depth Psychology. *Journal of Consciousness Studies* 9, no. 10: 11–36.

Throntveit, T. 2014. *William James and the Quest for an Ethical Republic*. New York: Palgrave Macmillan.

Uffelman, M. 2011. Forging the Self in the Stream of Experience: Classical Currents of Self-cultivation in James and Dewey. *Transactions of the Charles S. Peirce Society: A Quarterly Journal in American Philosophy* 47, no. 3: 319–339.

Varela, F. J. 1991. Organism: A Meshwork of Selfless Selves. In *Organism and the Origins of Self*, ed. A. I. Tauber, 79–107. Dordrecht, The Netherlands: Kluwer Academic Publishers.

Wiener, P. P. 1949. *Evolution and the Founders of Pragmatism*. Cambridge, MA: Harvard University Press.

Winther, R. G. 2014. James and Dewey on Abstraction. *The Pluralist* 9, no. 2: 1–28.

Winther, R. G. and S. Oyama. 2001. Selectional, Instructional and Maturational Theories in Evo-Devo and Behavior. Conference presentation, *International Society for the History, Philosophy, and Social Studies of Biology, 2001 Quinnipiac Session*.

1 Individuals in evolution

James's Darwinian psychology

The rise of scientific psychology coincided with the acceptance of evolutionary theories into mainstream science. Just as sensation, cognition and behaviour came to be studied as organic functions, the very concept of an organic function took on a radically historical dimension. As a result, evolutionists were now claiming that mental processes derive from a nervous system that had developed over millennia to manage the concrete exigencies of life. Mind was no longer an exalted faculty of knowledge but a set of instrumental functions tethered to an environment that had slowly ground them out.

Nineteenth-century evolutionism does not speak with one voice, however. The competing mechanisms of natural selection and the inheritance of acquired characteristics, for instance, suggest different relationships between ontogeny and phylogeny – individual development and the history of species – and each mechanism has been subject to competing interpretations that emphasize different internal and environmental factors. Tracing the influence of evolutionism in psychology therefore requires attending to the specific logics of particular evolutionary theories.

The present chapter does not attempt to map out this entire territory but instead traces one instructive path: that of William James. James was among the most influential psychologists in the decades following the publication of Darwin's *On the Origin of Species* (1859), especially in the US. Known as the Father of American Psychology, James is credited with founding the first psychology laboratory in the US in 1875;[1] teaching the first physiological psychology course in the US in the same year; supervising the country's first PhD in psychology in 1878; and publishing the seminal text of early US psychology, *The Principles of Psychology*, in 1890. An examination of James's scientific education and early writings show how he understood Darwinism to be emblematic of the uncertainty of science. James nevertheless embraced Darwin's theory, not just in explaining mental evolution, but also in modelling individual cognition and behaviour. James thus employs selectionist logics at both phylogenetic and ontogenetic levels, making him the first *double-barrelled Darwinian psychologist*.

This chapter outlines the sources of James's Darwinism in his education and early publications, before investigating his shift to an overtly Darwinian and anti-Lamarckian position in *The Principles of Psychology*. This investigation unearths an interesting interpretive tension. On the one hand, James presages neo-Darwinism in his physiological approach to mental life, his early repudiation of the inheritance of acquired characteristics and his creative extensions of Darwinian logic to non-biological domains. On the other hand, the twin lessons of Darwinian psychology for James are, first, that both science and philosophy are open-ended processes of fallible, inductive guesswork; and second, that the individual as such is a real locus of agency in the world. Darwinism for James signals a world that is both theoretically and actually *in the making*, with the individual as an active participant.

Uncertain science

Darwinism represents a distinctively modern kind of uncertain science. James learned this lesson from his teachers and intellectual companions at Harvard and expressed it in his very earliest publications.

An education in Darwinism

James matriculated at Harvard's Lawrence Scientific School in 1861, fast on the heels of the 1859 publication of Darwin's *Origin*. At Harvard James had contact with a series of luminaries who helped to define modern scientific education and research, all in a context where Darwinism was the defining issue of the time.[2] James first enrolled as a student of chemistry under Charles Eliot, a Darwinian who would oversee Harvard's transformation into a world-class research institution during his epic 40-year tenure as the university's president. Eliot instituted Harvard's elective system and heavily promoted the study of empirical science – two factors that worked in James's favour when Eliot eventually hired back his former student to teach niche courses on physiological psychology and evolution in the 1870s. The young James did not study with Eliot long, however, and quickly turned his attention from chemistry to anatomy. Here James studied under Jeffries Wyman. Although Wyman could not reconcile blind natural selection with his Unitarian theism, he accepted a form of spiritually guided evolution and provided James with formal training in Darwinism. James considered Wyman his exemplar of humble and disinterested scientific inquiry. This approach to science was shared by the great Harvard botanist Asa Gray, who did much to unify the taxonomy of plants in North America. Gray considered blind selection to be compatible with his theism, but he allowed a role for divine agency in the generation of variation.[3] James did not study with Gray directly, but he was aware of his presence and witnessed him speaking publicly in favour of Darwinism. These figures were more than just teachers. They were active researchers who were grappling with Darwin's new theory when its implications were just beginning to be understood. Indeed, both Wyman and Gray were personal correspondents of Darwin's, with whom they shared the details of their research.

The dominant scientist at Harvard when James arrived there was no Darwinian at all, however. On the contrary, it was stalwart anti-evolutionist Louis Agassiz. The addition of Agassiz to the faculty at Lawrence had been instrumental in the legitimization of the school as an institution and in the professionalization of the sciences more broadly in the US. Agassiz hailed from Switzerland and had studied with master French anatomist Georges Cuvier. He also had the distinction of being the first to propose a past global 'ice age' in his geological theory. From his authoritative post, Agassiz inveighed against evolutionism – and especially Darwinism – to the best of his ability. Agassiz and Gray were openly antagonistic on this question, and James witnessed one of Agassiz's lectures against Darwinism within months of beginning his studies.[4] James also got to know Agassiz better by joining him on an expedition to South America in 1865. One motivation for this trip for Agassiz and his financial backers was to find evidence undermining natural selection. James was taking Darwin's theory seriously by this point, however, and his opinions of Agassiz were decidedly mixed. As James writes to his father from the Amazon,

> I have profited a great deal by hearing Agassiz talk, not so much by what he says, for never did a man utter a greater amount of humbug, but by learning the way of feeling of such a vast practical engine as he is. *No one sees farther into a generalization than his own knowledge of details extends*, and you have a greater feeling of weight and solidity about the movement of Agassiz's mind, owing to the continual presence of this great background of special facts, than about the mind of any other man I know.[5]

James would remain convinced that generalizations are worth no more than the concrete experiences on which they are based, which is ultimately what 'cashes them out'. This is one sense in which James remained a lifelong empiricist. The irony of Agassiz is that, while he was oriented toward facts and taught in an aggressively hands-on manner, he did not really respect the right of facts to undermine his procrustean worldview. Species for Agassiz were static and ideal reflections of divine will. No amount of observation could have changed that axiom. Agassiz was thus a transitional figure in the history of biology who sided with the outgoing generation. If Gray and Wyman represented to James a cautiously modern approach to knowledge, then Agassiz represented an outmoded faith in idealized form. Wyman is 'the paragon ... of goodness, disinterestedness, and the single-minded love of truth',[6] while Agassiz's '*charlatanerie* is almost as great as his solid worth.... He wishes to be too omniscient'.[7]

Despite his contact with these towering figures, James's education meandered uncertainly. Indeed, James was a chronic vacillator who was often unsure about major life decisions. By the time of his South America trip, James had enrolled in Harvard Medical School under the supervision of Oliver Wendell Holmes, Sr. In this context, his South American excursion with Agassiz could be interpreted as an escape from uncertainties about his prospects as a physiologist or doctor. The same could be said of James's 1867/1868 trip to Germany, where he travelled to study the new field of experimental psychology in its homeland. James finally earned his medical degree after returning to Harvard in 1869. This was James's only earned degree, although he would never practice medicine. Instead, he took various appointments at Harvard in physiology, psychology and (eventually) philosophy. In James's words,

> I originally studied medicine in order to be a physiologist, but I drifted into psychology and philosophy from a sort of fatality. I never had any philosophic instruction, the first lecture on psychology I ever heard being the first I ever gave.[8]

The Metaphysical Club

James's education in evolutionary theory did not come only from his teachers. He also found an enviable cadre of intellectual companions. Of particular relevance is the Metaphysical Club, a discussion group that seems to have centred on Harvard in the 1870s.[9] In addition to James, this group included Charles Sanders Peirce, Chauncey Wright, John Fiske and Oliver Wendell Holmes, Jr., among others. The Metaphysical Club was a major incubator of American pragmatism, especially insofar as this school may be traced to Peirce and to James's adaptation and popularization of certain of his key ideas. The role of evolutionary theory in this group was pronounced. John Fiske was America's foremost promulgator of the evolutionary philosophy of Herbert Spencer, which he combined with his own optimistic brand of theism. Peirce was so taken by the notion of randomness begetting order that he generalized it into a cosmic principle of absolute chance or *tychism*. Chance for Peirce – unlike for Darwin – refers not to our ignorance but to an objective feature of the world. James would follow Peirce in this view and make it a cornerstone of his metaphysics of indeterminism.

By all accounts, however, the dominant member of the Metaphysical Club was Chauncey Wright. Wright was of a depressive bent and published very little. In James's words, he resembled 'more a philosopher of the antique or Socratic type than a modern *Gelehrter* [academic]'.[10] Nevertheless, he was a brilliant and forceful thinker. Like Fiske, Wright was an avowed positivist in the nineteenth-century sense of this term. That is, he was empirically oriented, sceptical about traditional philosophy and focussed on the logic of science.[11] Wright was also positivist in the sense of believing strongly in the value-neutrality of science and in spurned gratuitous metaphysical posits such as the scholastic concept of substance. Indeed, he coined the term 'cosmical weather' to describe the arbitrary aggregation and disaggregation of orderly patterns in the universe, which require no deep or rational explanation for their comings and goings.[12] He was also an early and vocal champion of Darwinism, who (like Wyman and Gray) numbered among Darwin's valued personal correspondents. In fact, he once published a defence of Darwinism against English biologist St George Jackson Mivart that Darwin so appreciated that he had it printed and circulated as a pamphlet in England at his own expense.[13]

James resisted Wright's denial of any metaphysical connectedness in the world, calling this view 'nihilism' and claiming that Wright makes of the universe a 'Nulliverse'.[14] On James's view, reality or experience natively contains connections as well as disconnections, even if there is no overarching absolute or rational connection holding the entire structure together. Assuming that disconnection is native and that connection must be imposed is, on James's view, a great mistake of traditional empiricism that distinguishes it from his own *radical* empiricism. Nevertheless, James did follow his positivistic sparring partner toward a more empirically grounded Darwinian outlook and away from the teleological Lamarckism of Fiske and Spencer.[15]

Juvenilia

James's early publishing record reflects the formative influence of Darwinism. His very first published pieces are two unsigned reviews of works of biology that he wrote while he was still a medical student in 1865.[16] Each of these reviews shows James cautiously supporting natural selection, while pondering the theory's philosophical implications.

The first review is of Thomas Henry Huxley's *On the Classification of Animals, and on the Vertebrate Skull*.[17] A pugilistic evolutionist, Huxley was also known as 'Darwin's Bulldog'. Although James's review is ostensibly a discussion of Huxley's contribution to the debate about whether the skull consists of modified vertebrae – Huxley thought not – James uses it mainly to examine the goals of science within the broad context of Darwinian biology. Thus, even in the midst of his physiological and medical training James was drawn toward broader philosophical questions about scientific logic and practice. Here James criticizes Cuvier, in a veiled criticism of Cuvier's student Agassiz:

> Now we are sure that biological Science, eternally grateful as she must be to Cuvier, will not consent to stop at these limits. Her function is not merely to note Resemblance, but to find Unity. Below the *fact* of resemblance she will seek till she lays bare the *ground* of resemblance; she will regard classification as her starting point rather than her goal; and far from spurning the 'System', she will proclaim that the creation of a perfect system is the very end of her existence. If Cuvier had lived two centuries earlier he would have been satisfied with knowing the coincidences that Kepler had discovered in the planetary orbits, and would have said, as Leibnitz actually did say, that Newton was impious to try to find their cause.[18]

For Cuvier as for Agassiz, the goal of anatomy is simply to limn the ideal structures of organic form. Such a science studies relationships of similarity and dissimilarity, but it will stop here lest it presume impiously to understand the divine plan. If science may legitimately seek a *natural ground* for organic form, however, then evolutionary theories would seem to provide just that. James thus recognizes the power of Darwinism to add a meaningful historical

dimension to science, as opposed to the rationalistic tendency simply to catalogue and admire.

James's second review is of Alfred Russell Wallace, whose co-discovery of
natural selection had forced a procrastinating Charles Darwin to go public with
his theory. James reviews Wallace's 1864 essay 'The Origin of Human Races'.
James agrees with Wallace that human races do not comprise distinct species.
He also agrees that the human species has ceased to evolve physically and will
henceforth evolve only mentally and socially. According to Wallace, humankind
has evolved a social and sympathetic nature through a form of group selection,
making mental and social variation the avenue of its further development. The
task is no longer to adapt to some particular environment but rather to become
ever smarter and more sympathetic so as to survive any given environment. Evolutionary specialization gives way to a kind of evolutionary generalization. In
James's words, 'The physical part of him is left immutable, and his mental and
moral advance is secured.'[19] The point is not that physical variation is no longer
generated or that mental changes do not depend upon changes within the body.
It is only that human evolution will no longer proceed through the selection of
gross physical morphology.

Soon after these first pieces, James wrote two separate reviews of Darwin's
Variation of Animals and Plants under Domestication (1868).[20] These reviews
demonstrate James's awareness of the uncertain and even dubious aspects of Darwin's theory. James focuses on Darwin's integrated theory of development, heredity and variation, which he called 'pangenesis'. On this view, microscopic units
called 'gemmules' direct the development of cells. Gemmules are collected into
the reproductive organs at certain stages of development so that they may be
passed on to offspring during reproduction. Once in the offspring, these units
direct the development of its own corresponding cells. According to Darwin, variation is introduced in this process by two types of cause: first, external impingements upon the organism; and second, internal physiological responses to these
impingements. James underscores the speculative nature of this theory in one his
reviews, especially regarding the mysterious internal causes of variation:

> The one strong impression that affects the reader, after laying down these
> volumes, is that of the endless complication of the phenomena in question,
> and the (perhaps hopeless) subtlety and occultness of the immediate causes.
> At the first glance, the only 'law' under which the greater mass of the facts
> the author has brought together can be grouped seems to be that of
> Caprice.[21]

This is a serious charge: An evolutionary theory without a viable theory of variation and heredity is incomplete. Where do new traits come from and how are
they entrenched in a population? Darwin harbours a dubious promissory note at
the heart of his theory.

These worries connect to broader concerns about probability and statistical
method. Natural selection was among the earliest scientific theories to rely

substantially on probabilistic evidence.[22] Darwin could not demonstrate with certainty that natural selection has been a significant factor in evolution, let alone that any particular trait was certain to have arisen or will arise in the future. He could only claim that with enough heritable variation – of the right kinds, with enough frequency – one could reasonably expect to see certain types of evolutionary change that we do in fact see. To anyone who associated scientific laws with certainty, natural selection looked like a sorry law, indeed. What is worse, natural selection required a longer geological timeframe than many in the nineteenth century were willing to accept. Blind natural selection acting upon accidental variation could not have produced complex life within a few thousand years. James addresses these worries in his other review of Darwin as follows:

> Perhaps from the very nature of the case, and the enormous spaces of time in question, it may never be any more possible to give a physically strict proof of [Darwin's theory], complete in every link, than it now is to give a logically binding disproof of it. This may or may not be a misfortune; at any rate it removes the matter from the jurisdiction of critics who are not zoölogists, but mere reasoners (and who have already written nonsense enough about it), and leaves it to the learned tact of experts, which alone is able to weigh delicate facts against each other, and to decide how many possibilities make a probability, and how many small probabilities make an almost certainty.[23]

James thus cites uncertainties about variation – whether there have been enough variational 'possibilities' to make it a 'probability' that natural selection explains extant organic forms – as reason for caution. In doing so, he positions himself alongside Wyman and Gray in viewing natural selection as a plausible hypothesis but not an indubitable truth.

Accepting natural selection would seem to require a particular attitude toward scientific theory-confirmation. As much as Darwin was a careful and observant naturalist, natural selection was not a law gleaned from successive observations in the manner of Baconian induction. The vast history of life is not susceptible to laboratory replication, nor can it be reduced to the billiards-style interactions of classical physics. Rather, to borrow a term from Peirce, Darwin seems to have employed a healthy dose of *abduction*. Abduction is also known as inference to the best explanation, or more colloquially, guessing. Guessing in this sense means observing a set of extant facts and proposing an explanation that implies or makes sense of them. This process is neither naively inductive nor strictly deductive. On the contrary, it allows a role for creativity, analogy and intuition. In Darwin's case, natural selection was a powerful guess – spurred by an analogy with selective breeding and a generalization of Thomas Malthus's social 'struggle for existence' – that made intelligible a set of disparate facts about anatomy and biogeography. Today we might regard this kind of guessing as the crucial hypothesis-formation moment of the hypothetico-deductive model, on which science proceeds through positing hypotheses and testing their deduced consequences in the world.[24] However, this understanding of science was only beginning to be theorized in the nineteenth century by figures like Peirce, William Whewell and William Stanley Jevons. To many it still looked suspiciously like putting the theoretical cart in front of the empirical horse.

Darwin's theory was uncertain in a deep way. Thus, even if Darwin's *Origin* opened the floodgates to the serious consideration of evolution within the mainstream scientific community, it did not lead to the immediate widespread acceptance of natural selection in particular. Instead, the cornerstone of Darwin's theory spent the end of the nineteenth century and beginning of the twentieth competing with an array of other proposed mechanisms. This included different forms of the inheritance of acquired characteristics, commonly known as 'Lamarckism' after its founder Jean-Baptiste Lamarck; the 'mutation theory' that explained evolution by way of discontinuous large-scale variations rather than continuous cumulative Darwinian ones; and the 'orthogenetic' idea that evolution is bound to unfold in preordained directions irrespective of environmental influence. Julian Huxley therefore dubbed this period the 'Eclipse of Darwinism' in his 1942 survey of evolutionary biology.[25] It was not until the 'Modern Synthesis' of natural selection with statistical population genetics in the 1930s and 1940s that the scientific community reached its current consensus that natural selection is a legitimate explanation of evolution and indeed the most important evolutionary factor.

Nevertheless, James embraced Darwinism just as it entered its phase of eclipse. Why?

Double-barrelled Darwinian psychology

James's embrace of Darwinism was never just about Darwin's theory of natural selection in the strictest sense. It was always about James's own promiscuous uses of Darwinian logic. For instance, James highlights an analogy that a number of theorists have remarked upon since: The theory of natural selection describes a process in nature that shares a pattern with guessing in individuals. The logic in both cases is that of trial and error, or generate-and-test. Natural selection is Darwin's guess that nature is a guesser. With the partial exception of Wright, James was the first to endorse such a doubly Darwinian perspective within the field of psychology.[26] James was thus the first *double-barrelled Darwinian psychologist*.

James's Darwinian psychology also dovetails with other areas of selectionist theory in his work. Indeed, James holds that the developing individual generates variation, not only for multiple interacting selectionist processes within ontogeny, but also for socio-historical and epistemological systems that fall outside of the traditional scope of psychology or biology. These systems are elaborated at more length in subsequent chapters, but it is worth noting here how they are firmly grounded in his psychological writings.

Selective perception

James's promiscuity with Darwinian logic can be traced at least as early as his 1875 review of Wilhelm Wundt's *Grundzüge der physiologischen Psychologie* (1874).[27] This piece is interesting if only as a review of the founder of the German laboratory tradition in psychology by an unknown writer who would, by the turn of the twentieth century, rise to become his counterpart in the US. Of particular note, however, is that it contains in germinal form two central ideas of James's mature Darwinian psychology. The first is that a Darwinian viewpoint provides prima facie reason to believe in the efficaciousness of consciousness. In James's words, 'unless consciousness served some useful purpose, it would not have been superadded to life'.[28] (This idea is examined in Chapter 3.) The second point is that mind itself is selective in a way that limits the power of the environment to determine cognition and behaviour.[29] James thus emphasizes the biasing power of what he calls 'interests':

> *My* experience is only what I agree to attend to. Pure sensation is the vague, a semi-chaos, the *whole* mass of impressions falling on any individual are chaotic, and become orderly only by selective attention and recognition. These acts postulate *interests* on the part of the subject, – interests which, as ends or purposes set by his emotional constitution, keep interfering with the pure flow of impressions and their association, and causing the vast majority of mere sensations to be ignored.[30]

If mind has evolved to be efficacious and to bias how the world is perceived, then neither the ontogeny nor phylogeny of mental life is wholly a process of coercive moulding by an external environment. Individuals make a difference.

James develops this construal of mind as selective environment in response to Herbert Spencer's *Principles of Psychology* (1855). The first work of modern evolutionary psychology, Spencer's book preceded Darwin's *Origin of Species* and embedded mind within a fully evolutionary system of nature.[31] Spencer posits a single integrated process of cosmic evolution that has physical, biological, psychological and social aspects. This integrated evolutionary process is characterized by a movement from the homogenous to the heterogeneous, or the simple to the complex. In the biological sphere, this tendency manifests itself as species becoming increasingly complex and differentiated in their morphology, cognition and behaviour. The mechanism of biological adaptation for Spencer was a version of the Lamarckian mechanism of the inheritance of acquired characteristics.[32] On Spencer's view, life itself consists in the 'continuous adjustment of internal relations to external relations' through Lamarckian means.[33] Mind, as an aspect of life, has its particular way of fulfilling this task, which is to 'correspond' to the environment.

Spencer's theory is significant in that it represents the precise meeting point of anglophone empiricist-associationist psychology and a more Continental tradition of positing rich innate mental structure. The former viewpoint is essentially externalist or 'outside-in', whereas the latter is essentially reliant on internal structures.

(Today this conflict endures in the form of disputes between cognitivists like Noam Chomsky and their detractors.) An evolutionary viewpoint allows these viewpoints to converge, or to moderate one another. By adding an evolutionary dimension to his empiricist account, Spencer avoids positing a mental 'blank slate' for any given individual by insisting that the slate had been inscribed over the generations. Innate structure is posited but is treated as a mutable historical product rather than a transcendental given. Some such 'evolutionary Kantianism' is still the going view in psychology today, even if we no longer prefer the Lamarckian mechanisms invoked by Spencer.

James takes issue with Spencer's account in his first signed scholarly essay, 'Remarks on Spencer's Definition of Mind as Correspondence' (1878). James believes that Spencer's view, while evolutionist, is still too externalist. This externalism suffuses his accounts of both ontogeny and phylogeny. According to Spencer, cognitive correspondence succeeds insofar as 'the persistence of the connexion between the states of consciousness is proportionate to the persistence of the connexion between the agencies to which they answer'.[34] Mental ontogeny is thus a reactive process of association, where strong connections in the world are met by strong connections in the mind. Mental phylogeny is then the cumulative Lamarckian total of such reactive processes over generations. James criticizes the passive nature of the knower on this view:

> I, for my part, cannot escape the consideration, forced upon me at every turn, that the knower is not simply a mirror floating with no foot-hold anywhere, and passively reflecting an order that he comes upon and simply finds existing. The knower is an actor, and co-efficient of truth on the one side, whilst on the other he registers the truth which he helps to create.... In other words, there belongs to mind, from its birth upward, a spontaneity, a vote. It is in the game, and not a mere looker on.[35]

In other words, James posits that interests are spontaneous in the same sense in which Darwinian variation is 'spontaneous': They arise from internal sources, and not for some environmental reason. James is explicit that he is drawing upon Darwinian logic, claiming that interests are 'incidentally implied in the workings of the nervous mechanism, and, therefore, in their ultimate origin, non-mental; for the idiosyncrasies of our nervous centers are mere "spontaneous variations", like any of those which form the ultimate *data* for Darwin's theory'.[36] Because interests are such brute irruptions, they are, according to James, 'the real a priori element in cognition'.[37]

This is a nuanced position: The mind generates 'spontaneous variation' and in this sense falls on the variation side of a variation-and-selection model. In doing so, however, it constitutes a biased environment for incoming stimuli, which it ignores or emphasizes based on factors endogenous to the mind. James thus posits two directions of mental selection: Mind is a *generator of variation* that it puts out in the world, as well as a *selective environment* for incoming sensory input.

Trial and error and the philosophy of science

James emphasizes the second direction in the above writings, construing the mind as selector of variation that bombards it from without. Nevertheless, the first direction represents a more traditional model of selectionism in psychology. In the context of learning theory, for instance, it gives us the idea of trial and error. Alexander Bain puts forth a physiological theory of trial-and-error learning in *The Emotions and the Will* (1859), first published in the same year as Darwin's theory of trial-and-error natural selection in *On the Origin of Species*. A reader of both Bain and Darwin, James was the first to combine these two levels of selection in a significant work of overtly Darwinian psychology, making him, as noted above, the first double-barrelled Darwinian psychologist.

Already in 1880, James had emphasized that the individual produces novel thoughts and behaviours that may or may not be reinforced by the environment. James believes that the human nervous system is a complex and 'instable' system. This system is rich in variation, which the external environment 'simply confirms or refutes, adopts or rejects, preserves or destroys – selects'.[38] Ideas and behaviours are thus produced internally and tried out against the world. Variation is relatively 'spontaneous' or non-directed in that it is not produced wholly in response to environment conditions, although it may become more directed as one learns. The seeming randomness of trial and error is not a disadvantage but an advantage. Without a range of possibilities on which to select, radically novel ideas do not get presented. A contribution either makes concrete facts better navigable or it does not, and its destiny as a 'truth' depends on its performance. Thus, '[t]o be fertile in hypotheses is the first requisite, and to be willing to throw them away the moment experience contradicts them is the next'.[39] In the *Principles*, James even includes a rudimentary theory of neurological reinforcement to show how learning is embodied.[40] Such a mechanism is analogous to a mechanism of heredity in phylogenetic evolution. That is, it shows how selections are preserved so that they may be reiterated in the future.

Interestingly, James goes beyond individual trial and error in outlining a selectionist model of the communal growth of scientific knowledge. One source for James's view is William Stanley Jevons's *Principles of Science* (1874). In reviewing the latter book, James takes particular note of the concept of trial and error in science:

> The account given of discovery as based wholly on this quick invention of hypothesis and subsequent verification is in the highest degree valuable and interesting, from the historic examples which illustrate it. Bacon's method of cataloguing instances and expecting the law passively to emerge at the end is shown to be ludicrously impotent, while Newton's and Faraday's practice of incessant guessing and testing are described as models.[41]

James especially rejects the notion that one could conceive of an idealization such as a natural law by passively observing regularities. The world does not inculcate us with knowledge by hitting us with it. The process requires an active contribution on the part of the knower. James makes this point using his characteristically Darwinian language, claiming that '[t]he conceiving of the law is a spontaneous variation in the strictest sense of the term'.[42] James writes to his colleague Hugo Münsterberg that their immature science of psychology is especially in need of this method: 'Whose *theories* in Psychology have any *definite* value today? No one's!... The man who throws out most new ideas and immediately seeks to subject them to experimental control is the most useful psychologist'.[43] Scientific knowledge-growth thus does not amount to the passive accumulation of extant facts. It is more like repeatedly seeking a foothold on the face of a mountain.

James nuances this view by positing that the selective environment for scientific hypotheses has multiple layers: the observed world, one's own holistic body of beliefs and the social world of peers, journals and the public. James again employs his Darwinian concept of selection as preservation:

> I read, write, experiment, consult experts. Everything corroborates my notion, which being then published in a book spreads from review to review and from mouth to mouth, till at least there is no doubt I am enshrined in the Pantheon of the great diviners of nature's ways. The environment *preserves* the conception which it was unable to *produce* in any brain less idiosyncratic than my own.[44]

These remarks indicate the hierarchical nature of James's selectionism, which has scarcely been touched on so far: Individuals generate variation that feeds any number of biological, psychological and social systems that radiate outward from them.

Trial-and-error learning theory lived on through James's student Edward Thorndike. Thorndike posited the 'Law of Effect', which is the simple idea that behaviours with satisfying consequences are likely to be repeated.[45] The Law of Effect was a precursor to behaviourist psychology's model of operant conditioning. This would seem to set the stage for behaviourists to follow James in proffering a double-barrelled Darwinian psychology, where trial and error is built upon an inherited structure that is itself explained in Darwinian terms. However, behaviourists like John Watson and B. F. Skinner actively downplayed the idea of an inherited mental endowment. Darwinian nativism was a threat to behaviourism's investment in high levels of behavioural plasticity: The more Darwinism explains, the less behaviourism does. Extreme behaviourism issues in a paternalistic and control-oriented management mindset, as in Skinner's disturbing book *Walden Two* (1948). This is not James's concept of trial-and-error learning as one tool in a pluralistic toolkit. On the contrary, James's career was forged in resistance to views that idolize a single theory or direction of causation, especially where this steamrolls the autogenous idiosyncrasies of individuals.[46]

James's view also presages more formalized theories of scientific knowledge-growth such as Karl Popper's.[47] Popper's falsificationism is a response to the problem of induction as underscored by David Hume: No amount of empirical evidence constitutes a certain or universal confirmation of a theory. The sun truly may not rise tomorrow. However, as represented by the logical rule *modus tollens*, we can at least *falsify* empirical generalizations by observing contradictions of a theory's entailments. This is the form that trial and error takes in Popper's view: We may test a theory, rule it out and let the remaining theories stand. Popper also famously considers falsifiability along these lines to constitute a demarcation criterion for what counts as science. James's view differs from Popper's in at least two ways: First, James's mature theory of truth ultimately focuses on verification rather than falsification, even if verification comes in degrees and produces no certainties; and second, James is less interested in demarcating science from non-science than he is in embedding science within the broader human endeavour.

James might have more affinities to Popper's anarchistic student Paul Feyerabend, who supported wide variation in scientific ideas and methods.[48] There is something of this flavour in James's celebration of the full breadth of human intellectual variation. James's selectionist model of science and later theory of truth also prefigure the work of twentieth-century social scientist Donald Campbell, who opened his 1977 William James Lecture at Harvard by quoting James on the variability of ideas produced by the spontaneous nervous system.[49]

Mental phylogeny

The preceding discussion of James's views on scientific knowledge-growth shows one sense in which he was doing what is now termed 'evolutionary epistemology'.[50] This is the modelling of knowledge as an evolutionary process unto itself, typically understood on analogy with natural selection. James also does evolutionary epistemology in a second sense, however, where the term refers to the study of knowledge in light of the evolution of cognition. These two positions are logically independent: One may model individual or communal knowledge-growth as a selection process while rejecting evolutionary explanations of cognitive structure; and one may accept the latter strategy while rejecting the former. As it happens, however, James was a selectionist pioneer in both areas.

The best source for James's position on mental phylogeny is *The Principles of Psychology* (1890), which can be viewed as a replacement for Spencer's Lamarckian work of the same name. A massive survey of the burgeoning science of mind, the *Principles* is distinguished from prior psychologies by its avowedly 'positivist' methodology. As James outlines in his preface, this means eschewing unexplained explainers like Christian souls and Cartesian or Kantian egos. The mind of the *Principles* needs no hidden metaphysical lynchpin, just as the cosmical weather of Chauncey Wright needs no deep explanation of its vanishing and arising forms. James instead seeks natural causes by adopting a physiological and evolutionary approach. James's *Principles* is in this way a distinctively modern psychological text.

Evolutionary debates play no small role in the book. In fact, James devotes the entire final chapter of the book to the examination of 'psychogenesis', or the evolution of inherited cognitive structure. James begins by rejecting two traditional explanations of the structure of mind: the empiricist's blank slate that is inscribed by experience, which provides no resources for even beginning to make sense of a world; and the neo-Kantian's transcendental conditions for the possibility of experience, which are inconsistent with James's naturalistic methodology and ignore the factual contexts of cognition.[51] As an alternative, James demands a *naturalized apriorism*, or a naturalistic account of inherited mental structure – in other words, just the sort of view pioneered by Spencer.

Spencer's account is not the only one available, however. James thus outlines two modes of explanation. The first is the way of 'adaptation', or Spencerian Lamarckism, whereas the second is the way of 'accidental variation', or Darwinian natural selection.[52] A classic illustration is the giraffe: One may explain the giraffe's long neck in terms of the cumulative inherited stretchings of each individual; one may also explain it in terms of the increased survival and reproduction of those individuals lucky enough to have longer necks than average. On the first (Lamarckian-instructionist) account, the species evolves because all of the individuals are changing in parallel. On the second (Darwinian-selectionist) account, the species evolves because of the shifting composition of the population over time.[53] James's use of the term 'adaptation' for the first view might be confusing, since today this term refers to the process (or products) of natural selection. Here 'adaptation' – with its connotations of passivity – is meant to capture the haplessness of organisms in Spencer's formulation of Lamarckism. To *adapt* is to be forced into shape and to pass on this shape to one's young.

James had clearly outlined the distinction between these views already in his 1880 essay 'Great Men, Great Thoughts, and the Environment'. James starts from the simple point that not everything in nature is directly relevant for explaining everything else. Most of nature can safely be ignored in any given explanation. Difficulties can arise in determining what is relevant to what, however, and just such a difficulty had obstructed evolutionary theory prior to Darwin: 'Predarwinian philosophers', unfortunately, 'all committed the blunder of clumping the two cycles of causation into one'.[54] Darwin avoids collapsing disparate cycles by positing a two-step process in which organisms offer up something (variation) for the environment to accept or reject (selection). This is Darwin's greatest achievement, in contradistinction to Lamarck and his followers.[55]

These two kinds of explanation are compatible in principle. Evolution is a complex process with a variety of factors, and both natural selection and the inheritance of acquired characteristics could be among those factors. Darwin himself never repudiated the inheritance of acquired characteristics. On the contrary, he made increased use of the latter idea in his later writings, especially *The Expression of the Emotions in Man and Animals* (1872). Similarly, Spencer made some room for natural selection, which he wryly dubbed 'the survival of the fittest' but preferred to call 'indirect equilibration' in his own work.[56] James himself took both views seriously until the late 1880s. By the time he was finishing the *Principles*, James believed strongly in Darwin's heritable non-directed variation.[57] He was still investigating Lamarckism, however, as in an 1888 letter to *Forest and Stream* in which he invites readers to contact him if they have evidence of the inheritance of acquired behavioural responses in hunting dogs.[58]

James is sceptical of Spencer's Lamarckian psychology because it does not answer the tough question that must be asked of any empiricism: How do we come to know anything at all if there is no prior structure to make sensation intelligible? It is true that Spencer attributes to each individual a rich innate cognitive structure that builds on the gains in cognitive 'correspondence' secured by past generations. However, he has only a vague conception of how this structure originated. Our ancestors could not have begun perceiving a world of related objects simply by being ensconced in it.[59] Spencer thus defers the problem of the blank slate into the mists of the evolutionary past. In this way he avoids the problem at the proximal level but not at the ultimate level. He can always invoke the necessary progressive increase of heterogeneity and correspondence, but this is ultimately a facile posit within an essentially deductive philosophy of nature.[60]

James understands that what is needed here is provided by natural selection but not Spencer's Lamarckism: non-directed variation. If Spencer holds that mind *had to arise* to make organisms better at corresponding, then James holds that mind could only have arisen *for no good reason*. A basic perceptual apparatus must first have been implemented, not out of an inherent organic tendency to correspond, but through dumb luck. This means that chance begets order in a genuine sense, where the order was not implicit in the prior inchoate state of affairs. From a cosmic perspective, this might make life and mind appear less rational and more arbitrary. However, value for James derives not from pedigree but from the mediation of possibilities for some purpose. Possibilities – whether in the development of a species, an individual or a society – are value-neutral until they go out and do some work.

James could stop with the conclusion that Spencer must give a necessary role to natural selection in his mostly Lamarckian view. He goes further, however, casting doubt upon the inheritance of acquired characteristics in general. Here James acknowledges the intuitive power of Spencer's argument in *The Factors of Organic Evolution* (1887) that organisms are too highly structured to be the result of mere chance and selection. Nevertheless, he argues that Spencer underestimates natural selection and that the inheritance of acquired characteristics is in any event empirically dubious.[61] The latter opinion owes crucially to James's then recent discovery of writings by German biologist August Weismann. Weismann's theory of heredity sequesters the germ cells from other bodily ('somatic') cells early in development such that the latter cannot affect the former in a systematic adaptive fashion.[62] As a result, Weismann's view rules out Lamarckism in principle. James does not positively endorse Weismann's theory, but he does think that it makes the German a particularly good critic of Spencer and Lamarck.[63] James also cites Weismann's laboratory experiments, the most famous of which involved removing the tails from successive generations of mice whose offspring grew tails nonetheless.[64] Weismann in this way buttresses

James's conviction that natural selection can explain any trait that is purportedly explained by the inheritance of acquired characteristics.

James thus concludes that 'the so-called Experience-philosophy' – that is, Spencer's Lamarckian empiricism – 'has failed to prove its point'.[65] This leaves James with a non-Lamarckian version of Darwinism, which is just the kind of stance that would sweep through biology in the decades following James's book.

Darwinism without neo-Darwinism

It is easy to see how James could figure as a hero in some neo-Darwinian Whig history. In the midst of the Eclipse of Darwinism, the hard-nosed physiologist James writes the first positivistic and Darwinian work of psychology. Here he pushes aside hoary metaphysical and Lamarckian theorists to inaugurate a new era of Darwinism in behavioural science. For good measure, he cleverly promulgates trial-and-error principles that would become central in experimental psychology and the philosophy of science. It is just icing on the cake that James cites Weismann, whose theory of heredity is considered the most important conceptual ancestor of the 'central dogma' of twentieth-century genetics – that is, the anti-Lamarckian view that information flows from genes to ribonucleic acid and proteins, but never back in the opposite direction to inform the genetic structure inherited by the next generation.

This impression should be tempered in a way that respects the seminal nature of James's work while also recognizing his historical context and uniqueness as a thinker. Several remarks can be made along these lines.

Labels and logics

The first point is that not too much should be made of James's 'positivism'. As suggested above, this term had a broader meaning in the nineteenth century and was adopted by many writers who were sceptical of traditional metaphysics and interested in the logic of science. Indeed, James was not much of a positivist in this *nineteenth-century* sense. He did argue that philosophy should be viewed as a completer and unifier of the sciences (including mental and moral science).[66] However, he never challenged the validity of metaphysical speculation, whether in the *Principles* or elsewhere. On the contrary, he openly rejected such an attitude, claiming that '[m]etaphysics of some sort there must be. The only alternative is between the good Metaphysics of clear-headed Philosophy and the trashy Metaphysics of vulgar Positivism'.[67] The fact is that James was only methodologically a positivist, and only in one book. He is explicit about this in his preface to the *Principles*, claiming that metaphysics is a legitimate pursuit but that an immature science should bracket such matters while getting its bearings.[68] Subjects such as metaphysics and ethics that are normally decentred by positivism are actually central to James's thought. James does formulate a *pragmatic method* to determine the meaning of metaphysical disputes, but his aim is not generally to dissolve them away.

Second, James plainly lacked much of the conceptual apparatus of later neo-Darwinism. Like Darwin, he did not hold a Mendelian or 'particulate' genetic theory. Nor did he hold a transmissional view of heredity, where inheritance consists in the transfer of the same unchanged genes with which one was born. Rather, he likely assumed the common nineteenth-century view that variation is introduced by way of impingements on hereditary processes during development.[69] Darwin's theory of pangenesis, mentioned above, is clearly a developmental view. Heredity on this theory is an ongoing developmental process, where 'a variation' occurs when this process is knocked off track.[70] (Compare this to 'variation' in the modern statistical sense, which refers not to countable traits but to aggregate information about differences within a population.) Even Weismann, who is lionized for anticipating a transmissional view, held that variations are induced in germinal materials during ontogeny.[71] Such a view does not imply Lamarckism in the familiar sense, as long as the impingements are not instructionist – that is, as long as they do not systematically impose heritable adaptive variation. As it happens, Darwin became increasingly Lamarckian in this instructionist sense, while James joined Weismann in rejecting such a view.

James is also distinguished by his positing of variation that is not only non-directed in relation to the environment but also *literally random*. James ultimately holds that the world is constructed through the mediation of random variation at multiple levels of reality, where the ultimate origins of variation are inscrutable. This view becomes clearer in James's later metaphysical writings (discussed in Chapter 5), which adapt elements from Peirce's doctrines of *tychism* and *synechism*. This may resemble certain interpretations of quantum mechanics,[72] but it is not a standard feature of neo-Darwinism.

Finally, James appears even further from neo-Darwinism when his motives are considered. Darwinism for James is principally a weapon against externalism, or the explanation of a system in terms of properties external to that system. James's combats Spencer's Lamarckian externalism by driving a Darwinian logical wedge between variation and selection, thus denying that mental ontogeny or phylogeny could consist wholly in the shaping of mind to fit a coercive world. This logical move affords James the idea of autogenous mental and behavioural variation, as well as the idea that the individual alters its own surrounding world in a manner that was not pre-programmed from without. It is important to recognize that Darwinism need not be conceived like this. It is more common today to conceive of natural selection as a strongly externalist force that sorts and winnows disembodied genes. Agency is attributed to the selective environment, which sets the problems that populations need to solve. It is not attributed to the spontaneous powers in the individual that are James's focus.

This externalist construal of natural selection follows an externalist construal of selectionism more broadly. Ron Amundson encapsulates this externalism in his definition of selectionism in terms of the following (paraphrased) criteria. According to Amundson, an explanation is 'forcefully' selectionist insofar as these conditions are met:[73]

- Richness of variation: wide variation, continually produced, with fine-grained differences among variants.
- Non-directedness of variation: variation that is not produced in a manner that biases it toward having a certain use.
- Non-purposive sorting mechanism: a selective environment that does not serve some purpose or goal.

These conditions are externalist because they are designed to give maximal explanatory credit to the selective environment. If there is a robust thicket of undifferentiated variation continually growing in all directions, then the environment's pruning will receive credit for whatever shape emerges. Without the environment there would only be an unstructured spherical growth.

Selectionist explanation at its ideal limit is thus asymmetrically externalist. This is reason to be cautious about identifying James's position too strongly with today's enthusiasts for selectionism. It is better to say that James places great emphasis on one particular principle that is definitive of selectionism: the non-directedness of variation. Indeed, emphasizing the variation side of a selectionist account is one way of challenging externalism: The environment can ratify or veto, but it cannot draft the legislation.[74] This is not the only way to challenge externalism, however. One can also posit internally driven variation within an instructionist account. A voluntarist Lamarckism, for instance, could posit that individuals purposively will themselves into incorporating desired characteristics, which they then pass on to future generations. Such a view may be far afield from Spencer – not to mention Lamarck, who was a wholly mechanistic thinker – but it has been attractive to some precisely because it seems more humane than the Darwinian image of brutal mechanistic winnowing. Both Darwinism and Lamarckism therefore have different formulations emphasizing different internal and external factors. One cannot infer much from a label.

Beyond Darwinism and Lamarckism

Selectionist and instructionist accounts are similar in that they give some important role to the environment. These are not the only types of account that can be invoked to explain features of complex systems. There are also *internalist* or internal-drive explanations, which tend to limit the explanatory power of instructionist or selectionist explanations (including natural selection in neo-Darwinism).[75] James puts forth several such explanations, adding further nuance to his conception of Darwinian evolution.

One such explanation is the '*Bauplan*' view that certain basic body types, once entrenched, constrain the possibilities for a lineage's further evolution. This concept gives James a sceptical attitude toward specific adaptationist hypotheses, to the point where he concludes *The Principles of Psychology* with the following word of caution about mental evolution. While mental structures are surely natural products, James contends that they *have all grown up in ways of which at present we can give no account.* Even in the clearest parts of our Psychology our insight is insignificant enough. And the more sincerely one seeks to trace the actual course of *psychogenesis*, the steps by which as a race we may have come by the peculiar mental attributes which we possess, the more clearly one perceives 'the slowly gathering twilight close in utter night.'[76]

James is thus not sanguine about our ability to infer evolutionary function for a given trait. For instance, he liked to repeat the claim that humans have five fingers 'merely because the first vertebrate above the fishes *happened* to have that number'.[77] James here prefigures later critiques of adaptationism, including most famously those levelled by Stephen Jay Gould and Richard Lewontin.[78] An adaptively neutral trait may tag along only because it is a corollary of some other trait. In fact, what looks like a well-defined trait might have no evolutionary significance at all.

Another perspective is represented by the *self-organization* view, championed by Stuart Kauffman, which gives significant evolutionary credit to universal laws governing the basic structure and dynamics of matter.[79] Already in his first book review, James makes a suggestive remark along these lines regarding Huxley's denial that skull segments are modified vertebrae:

> We think that the undeniable analogy of these segments to true vertebrae will some day be shown to be a true affinity. Both backbone and skull will be affiliated upon some uniform mode of force (working in either under slightly different conditions), in accordance with the principles of a synthesis which is now slowly shaping itself in biology. This synthesis asserts that organic forms, like the forms of the waves of the sea, are the result of the common properties of matter.[80]

James thus suggests that the analogous properties of different organic structures might be explained less by parallel instances of historical environmental shaping and more by invariant laws governing the physico-chemical structure of the world. This view may be termed 'internalist' or perhaps 'everywhere-ist' in that it invokes highly general regularities that pervade all parts of the system under consideration. In this way it may take away some credit from the environment considered as a special distinct force.

Finally, James is encouraged by internalist *saltationism*, which posits large-scale internally generated mutations. This is clear in his review of William Bateson's *Materials for the Study of Variation* (1894). Here Bateson challenges Darwin and Wallace's view that 'the discrete differences between existing species are due to the summation, pursued through successive generations, of numerous small variations in the same direction', while arguing that large-scale discontinuous mutations are the more important factor.[81] James is particularly interested in the potential of Bateson's theory to undercut externalist views of mental ontogeny and phylogeny:

As regards psychology, it is clear that the triumph of views like Mr. Bateson's will strengthen the hands of the anti-associationists, and in general of those who have contended for an autogenous origin of certain human faculties, of certain instincts and tastes, for example, or of conscience, the higher reason and the religious sense. The book is a masterly production, and unquestionably inaugurates a new department of research.[82]

Such a view gives the primary credit to internal systems, which produce new structures without the help of slow processes of environmental winnowing. James takes this view seriously, including Bateson's contention that different traits may co-mutate simultaneously to produce an entire new integrated system.

Each of these internalist views acts as a corrective to the inescapably externalist aspects of Darwin's theory. If the *Bauplan* and self-organization views share an emphasis on the constrained and predictable nature of evolution, saltationism is about how nature can deliver large-scale surprises unannounced. James's pluralistic willingness to welcome these viewpoints into his post-Darwinian worldview demonstrates that his allegiance lies with anti-externalism as a general explanatory strategy, not with formulating Darwin's theory in the most 'forceful' (externalist) way possible.

Conclusion

Darwinism provided James with an example of uncertain science. Rather than discarding this science, James took it to be paradigmatic of all of knowledge: fallible, provisional and verifiable in an abductive and holistic fashion. Curiously, Darwin's theory describes a process in nature that is structurally analogous to the process that generates scientific theories in the first place: trial and error. Natural selection has evolved a nervous system that acts as a scaffold for higher orders of selectionist processes occurring within each individual's life. Phylogenetic selection gives rise to ontogenetic selection, like the fractal reiteration of a geometrical shape.

James posits selectionist systems in mental phylogeny and ontogeny, making him the first double-barrelled Darwinian psychologist. If anything, James places more emphasis on ontogeny than phylogeny, where he posits two directions of psychological selection: first, an 'outside-in' direction where the mind acts as selective environment for incoming sensory stimuli; and second, an 'inside-out' direction where the mind generates behavioural variation that is selected by the external physical and social world. Both of these directions cut against Spencer's externalist Lamarckism, which portrays the individual as passive clay. James also takes Spencer to task for his inability to explain how mental phylogeny could have begun without variation that arose for no particular environmental reason.

It is perhaps unorthodox to describe all of this variation in Darwinian terms. We do not normally describe behavioural or cognitive oddities in ontogeny in the same way that we describe congenital Darwinian mutations. For James, however, it is the non-directedness of variation in *both* ontogeny and phylogeny that defines a Darwinian worldview. Congenital, genetically heritable variation may be an interesting subset of variation, but even non-heritable ontogenetic variation may have phylogenetic effects. This fact is captured, for instance, by the concept of the 'Baldwin Effect', or the idea – conceived roughly simultaneously by James Mark Baldwin and others around 1896 – that organisms may learn new behaviours that later become entrenched through natural selection.[83] Here ontogeny 'sets up' phylogeny, creating an effect that appears Lamarckian but is strictly compatible with neo-Darwinism. James does not seem to have commented upon the latter idea in print, although he knew Baldwin and was interested in ascribing a role to developing individuals in the systems that embed them. Nevertheless, James goes beyond such intra-biological debates to show how ontogenetic variation may influence systems such as the evolution of scientific knowledge and of society more broadly. As will become clearer in Chapter 2, not everything evolutionary is about biology.

Notes

1 This honour is sometimes bestowed on James's student G. Stanley Hall for his laboratory at Johns Hopkins. The issue turns on the level of research and sophistication required for something to count as a 'laboratory'. James once responded indignantly to a piece in Hall's own *American Journal of Psychology* that erroneously described the conditions of the Harvard lab's founding (ECR, 150). See also Bjork (1983).
2 For James's development in terms of the influence of his teachers, see Croce (1995a).
3 Gray 1876.
4 To his family in September of 1861, James writes: 'Agassiz gives now a course of lectures in Boston, to which I have been. He is evidently a great favorite with the audience and feels so himself' (LWJ I, 34–35).
5 LWJ I, 65. Emphasis added.
6 ECR, 9.
7 Quoted in Croce (1995b, 143).
8 Perry 1935 I, 228.
9 Menand 2002; Richardson 2006.
10 ECR, 15. The use of a German term is meant to convey an especial dryness and technicality. See also James's claim that the technical methods of modern psychology 'could hardly have arisen in a country whose natives could be *bored*' (PP I, 192).
11 Pearce 2015.
12 Wright 1864.
13 Wright 1871. This reprinted critical review was of Mivart's *On the Genesis of Species* (1871).
14 MEN, 154.
15 Peirce joined Wright in chiding James for his enthusiasm about Spencer. James refers to a time when he was a subscriber to Spencer's philosophy and thus received it serially ('in numbers'):

> I read [Spencer's *First Principles*] as a youth when it was still appearing in numbers, and was carried away with enthusiasm by the intellectual perspectives which it seemed to open. When a maturer companion, Mr. Charles S. Peirce, attacked it in my presence, I felt spiritually wounded, as by the defacement of a sacred image or picture, though I could not verbally defend it against his criticisms.
>
> (EPH, 116)

16 These pieces were not widely known until several decades after James's death, leaving early interpreters unaware of the full extent of Darwin's imprint on James. They are collected in *Essays, Comments, and Reviews* (1987).

17 ECR, 197–205.

18 ECR, 202.

19 ECR, 208.

20 James also discusses Darwinism in an 1868 review of French anthropologist Armand de Quatrefages (ECR, 216–221).

21 ECR, 234.

22 Krüger *et al.* 1987; Krüger *et al.* 1990; Croce 1995b; Menand 2002.

23 ECR, 239.

24 Ghiselin (1969) champions Darwinism as a benchmark in the development of the hypothetico-deductive method; Ayala (2009) argues that Darwin employed this method despite his attempts to portray his method as simply inductive; and Sober (2009) points out that 'hypothetico-*deductive*' is a misleading term, given the probabilistic nature of Darwin's theory.

25 Huxley 1942. Bowler (2003) makes much of the idea of such an eclipse.

26 Wright 1873. James made more extensive use of the logic of selection than Wright did, and he exerted more influence over both psychology and philosophy.

27 ECR, 296–303.

28 ECR, 302.

29 James found this point in Wundt, Wright and English philosopher Shadworth Hodgson. See Pearce (forthcoming).

30 ECR, 300.

31 James worked from the expanded second edition of 1872, which was more widely available in the US than the first. It also gained more attention due to the intervening controversies over Darwin's 1859 book.

32 Spencer found Lamarckism in Robert Chambers's popular work *Vestiges of the Natural History of Creation* and by reading a critical account in Charles Lyell's *Principles of Geology*.

33 Spencer 1872/1855 I, §131.

34 Spencer 1872/1855 I, §183. Emphasis removed.

35 EPH, 21.

36 EPH, 19.

37 EPH, 11n1.

38 WB, 640–641.

39 WB, 185.

40 PP II, 1186–1187.

41 ECR, 290.

42 WB, 186. See also PP II, 1232.

43 LWJ II, 312–313.

44 WB, 186.

45 Thorndike 1898.

46 Dennett (1975) also gives pride of place to the Law of Effect, claiming that it must be assumed a priori by psychology. See the Conclusion of the present study.

47 Popper 1934.

48 Feyerabend 1975.

49 Campbell 1988, 435–436. Campbell's (1974) evolutionary epistemology is like James's theory of truth in that it is based on selectionist principles and is essentially descriptive rather than explanatory. Campbell recognizes both Jevons and James as precursors.

50 Hahlweg and Hooker 1989.

51 As James later remarked, referring to the 'theory of knowledge' favoured by his contemporary neo-Kantians, 'the whole hocus-pocus of *erkenntnisstheorie* begins, and goes on unrestrained by further concrete considerations' (MT, 81).

52 PP II, 1224.
53 Sober 1984, 149.
54 WB, 167–168.
55 WB, 167. James was of course aware of Darwin's attempts to explain variation, which he had examined in multiple book reviews and which did include explanations in terms of external factors. James's point, however, is that natural selection *need not* depend on any direct systematic relationship between environment and the production of adaptive variation.
56 Spencer relegates natural selection to a footnote in the 1872 edition of *The Principles of Psychology*, claiming 'survival of the fittest to be always a co-operating cause' in evolution (§189). See also Spencer's *Principles of Biology* (1864, §166).
57 'The evidence for Mr. Darwin's view is too complex to be given in this place. To my mind it is quite convincing' (PP II, 1275).
58 ECR, 127–128.
59 PP II, 1226–1227.
60 James emphasized the vagueness of Spencer's evolutionism in a course titled 'The Philosophy of Evolution' that he taught repeatedly throughout the 1880s and 1890s. Here he parodied Spencer's position (and prose) as follows: 'Evolution is a change from a no-howish untalkable all-alikeness to a some-howish and in general talka-boutable not-all-alikeness by continuous sticktogetherations and somethingelseifica-tions' (quoted in Perry 1935 I, 482).
61 PP II, 1279.
62 PP II, 1278. James cites Weismann (1883, 1885).
63 PP II, 1278.
64 See also the ongoing experiment wherein societies have observed that each generation needs to be circumcised anew. To be fair, these observations only count against the inheritance of passively acquired alterations (such as injuries), not the Lamarckian 'use-inheritance' theory on which a trait is passed on because of the active engage-ment of some function.
65 PP II, 1280.
66 SPP, 19. Philosophy for James embeds and enlarges upon positive science.
67 EPH, 57.
68 PP I, 6.
69 Churchill 1987.
70 Winther 2000.
71 Winther 2001.
72 Stapp 2007.
73 Amundson 1989, 417.
74 The social implications of this view are spelled out in more detail in Chapter 2.
75 Winther and Oyama 2001.
76 PP II, 1280. Emphasis added. The final eight words are an uncited quotation from Book VII of Wordsworth's *The Excursion*.
77 WB, 178–179.
78 Gould and Lewontin 1979.
79 Kauffman 1995.
80 ECR, 202.
81 ECR, 498.
82 ECR, 500. For James as a saltationist, see also Godfrey-Smith (1996, 91–93).
83 Baldwin 1896; Weber and Depew 2003.

References

Amundson, R. 1989. The Trials and Tribulations of Selectionist Explanations. In *Issues in Evolutionary Epistemology*, eds K. Hahlweg and C. A. Hooker, 413–432. Albany, NY: State University of New York Press.

Ayala, F. J. 2009. Darwin and the Scientific Method. *Proceedings of the National Academy of Sciences* 106, Supplement 1: 10,033–10,039.

Bain, A. 1859. *The Emotions and the Will*. London: Parker.

Baldwin, J. M. 1896. A New Factor in Evolution. *The American Naturalist* 30, no. 354: 441–451.

Bateson, W. 1894. *Materials for the Study of Variation: Treated with Especial Regard to Discontinuity in the Origin of Species*. Cambridge: Cambridge University Press.

Bjork, D. W. 1983. *The Compromised Scientist: William James in the Development of American Psychology*. New York: Columbia University Press.

Bowler, P. J. 2003. *Evolution: The History of an Idea*. 3rd edn. Berkeley, CA: University of California Press.

Campbell, D. T. 1974. Evolutionary Epistemology. In *The Philosophy of Karl Popper*, ed. P. A. Schilpp, 413–463. La Salle, IL: Open Court.

Campbell, D. T. 1988. Descriptive Epistemology. In *Methodology and Epistemology for Social Science: Selected Papers*, 435–486. Chicago, IL: University of Chicago Press.

Chambers, R. 1844. *Vestiges of the Natural History of Creation*. London: J. Churchill.

Churchill, F. B. 1987. From Heredity Theory to Vererbung: The Transmission Problem, 1850–1915. *Isis* 78, no. 3: 337–364.

Croce, P. J. 1995a. William James's Scientific Education. *History of the Human Sciences* 8, no. 1: 9–27.

Croce, P. J. 1995b. *Science and Religion in the Era of William James: Volume 1, Eclipse of Certainty, 1820–1880*. Chapel Hill, NC: University of North Carolina Press.

Darwin, C. 1859. *On the Origin of Species By Means of Natural Selection, Or The Preservation of Favoured Races in the Struggle for Life*. London: John Murray.

Darwin, C. 1868. *The Variation of Animals and Plants under Domestication*. 2 vols. London: John Murray.

Darwin, C. 1872. *The Expression of the Emotions in Man and Animals*. London: John Murray.

Dennett, D. C. 1975. Why the Law of Effect Will Not Go Away. *Journal for the Theory of Social Behaviour* 5, no. 2: 169–188.

Feyerabend, P. 1975. *Against Method: Outline of an Anarchistic Theory of Knowledge*. London: Verso.

Ghiselin, M. T. 1969. *The Triumph of the Darwinian Method*. Berkeley, CA: University of California Press.

Godfrey-Smith, P. 1996. *Complexity and the Function of Mind in Nature*. Cambridge Studies in Philosophy and Biology. Cambridge: Cambridge University Press.

Gould, S. J. and R. C. Lewontin. 1979. The Spandrels of San Marco and the Panglossian Paradigm: A Critique of the Adaptationist Programme. *Proceedings of the Royal Society of London. Series B, Biological Sciences* 205, no. 1161: 581–598.

Gray, A. 1876. *Darwiniana*. New York: Appleton.

Hahlweg, K. and C. A. Hooker. 1989. *Issues in Evolutionary Epistemology*. State University of New York Series in Philosophy and Biology. Edited by D. E. Shaner. Albany, NY: State University of New York Press.

Huxley, J. 1942. *Evolution: The Modern Synthesis*. London: George Alien & Unwin Ltd.

Huxley, T. H. 1864. *Lectures on the Elements of Comparative Anatomy: On the Classification of Animals and on the Vertebrate Skull.* London: Churchill.

James, W. 1975. *The Meaning of Truth.* The Works of William James. Edited by F. Burkhardt, F. Bowers and I. K. Skrupskelis. Cambridge, MA: Harvard University Press. Original edition, 1909.

James, W. 1978. *Essays in Philosophy.* The Works of William James. Edited by F. Burkhardt, F. Bowers and I. K. Skrupskelis. Cambridge, MA: Harvard University Press.

James, W. 1979. *Some Problems of Philosophy.* The Works of William James. Edited by F. Burkhardt, F. Bowers and I. K. Skrupskelis. Cambridge, MA: Harvard University Press. Original edition, 1911.

James, W. 1979. *The Will to Believe and Other Essays in Popular Philosophy.* The Works of William James. Edited by F. Burkhardt, F. Bowers and I. K. Skrupskelis. Cambridge, MA: Harvard University Press. Original edition, 1897.

James, W. 1981. *The Principles of Psychology.* 2 vols. The Works of William James. Edited by F. Burkhardt, F. Bowers and I. K. Skrupskelis. Cambridge, MA: Harvard University Press. Original edition, 1890.

James, W. 1987. *Essays, Comments, and Reviews.* The Works of William James. Edited by F. Burkhardt, F. Bowers and I. K. Skrupskelis. Cambridge, MA: Harvard University Press.

James, W. 1988. *Manuscript Essays and Notes.* The Works of William James. Edited by F. Burkhardt, F. Bowers and I. K. Skrupskelis. Cambridge, MA: Harvard University Press.

James, W. and H. James. 1920. *The Letters of William James: Two Volumes Combined.* Boston, MA: Little, Brown, and Co.

Jevons, W. S. 1874. *The Principles of Science: A Treatise on Logic and Scientific Method.* London: Macmillan and Co.

Kauffman, S. A. 1995. *At Home in the Universe: The Search for Laws of Self-Organization and Complexity.* New York: Oxford University Press.

Krüger, L., L. Daston and M. Heidelberger, eds. 1987. *The Probabilistic Revolution. Volume 1: Ideas in History.* Cambridge, MA: MIT Press.

Krüger, L., G. Gigerenzer and M. Morgan, eds. 1990. *The Probabilistic Revolution. Volume 2: Ideas in the Sciences.* Cambridge, MA: MIT Press.

Lyell, C. 1837. *Principles of Geology: Being an Inquiry How Far the Former Changes of the Earth's Surface are Referable to Causes Now in Operation.* 4 vols. London: Murray.

Menand, L. 2002. *The Metaphysical Club: A Story of Ideas in America.* New York: Strauss and Giroux.

Mivart, St G. 1871. *On the Genesis of Species.* New York: D. Appleton.

Pearce, T. 2015. Science Organized: Positivism and the Metaphysical Club, 1865–1875. *Journal of the History of Ideas* 76, no. 3: 441–465.

Pearce, T. Forthcoming. James and Evolution. In *Oxford Handbook of William James*, ed. Alexander Klein. New York: Oxford University Press.

Perry, R. B. 1935. *The Thought and Character of William James.* 2 vols. Boston, MA: Little, Brown, and Company.

Popper, K. R. 1934. *Logik der Forschung: Zur Erkenntnistheorie der modernen Naturwissenschaft.* Schriften zur Wissenschaftlichen Weltauffassung. Edited by P. Frank and M. Schlick. Vienna: Verlag von Julius Spring.

Richardson, R. D. 2006. *William James in the Maelstrom of American Modernism: A Biography.* Boston, MA: Mariner Books.

Skinner, B. F. 1948. *Walden Two*. Indianapolis, IN: Hackett.

Sober, E. 1984. *The Nature of Selection*. Cambridge, MA: MIT Press.

Sober, E. 2009. Did Darwin Write the Origin Backwards? *Proceedings of the National Academy of Sciences* 106, Supplement 1: 10,048–10,055.

Spencer, H. 1864. *Principles of Biology*. 2 vols. London: Williams and Norgate.

Spencer, H. 1872. *The Principles of Psychology*. 2 vols. London: Longmans. Original edition, 1855.

Spencer, H. 1887. *The Factors of Organic Evolution*. London: Williams and Norgate.

Stapp, H. 2007. Whitehead, James, and the Ontology of Quantum Theory. *Mind and Matter* 5, no. 1: 83–109.

Thorndike, E. L. 1898. Animal Intelligence: An Experimental Study of the Associative Processes in Animals. PhD thesis. Columbia University, New York.

Wallace, A. R. 1864. The Origin of Human Races and the Antiquity of Man Deduced from the Theory of 'Natural Selection'. *Journal of the Anthropological Society of London* 2, clviii–clxxxvii.

Weber, B. H. and D. J. Depew, eds. 2003. *Evolution and Learning: The Baldwin Effect Reconsidered*. Cambridge, MA: MIT Press.

Weismann, A. 1883. *Über die Vererbung*. Jena, Germany: Gustav Fischer.

Weismann, A. 1885. *Die Continuitat des Keimplasma's als Grundlager einer Theorie der Vererbung*. Jena, Germany: Gustav Fischer.

Winther, R. G. 2000. Darwin on Variation and Heredity. *Journal of the History of Biology* 33, no. 3: 425–455.

Winther, R. G. 2001. August Weismann on Germ-Plasm Variation. *Journal of the History of Biology* 34, no. 3: 517–555.

Winther, R. G. and S. Oyama. 2001. Selectional, Instructional and Maturational Theories in Evo-Devo and Behavior. Conference presentation, *International Society for the History, Philosophy, and Social Studies of Biology, 2001 Quinnipiac Session*.

Wordsworth, W. 1814. *Excursion: Being a Portion of The Recluse, a Poem*. London: Longman, Hurst, Rees, Orme and Brown.

Wright, C. 1864. 'A Physical Theory of the Universe'. Review Essay on Herbert Spencer. *Essays: Scientific, Political, and Speculative. North American Review* 99, no. 1: 1–34.

Wright, C. 1871. *Darwinism: Being an Examination of Mr. St. George Mivart's 'Genesis of Species'*. London: John Murray.

Wright, C. 1873. Evolution of Self-Consciousness. *The North American Review* 116, no. 239: 245–310.

Wundt, W. 1874. *Grundzüge der physiologischen Psychologie*. Leipzig, Germany: Engelmann.

2 Individuals in history
Social evolution without social Darwinism

Evolution belongs to no discipline in particular. On the contrary, evolutionary theory arose through a complex dialectic among the physical, social and life sciences, over a period when the current academic disciplinary matrix had not yet been crystallized. A well-known example of this dialectic is the influence of Thomas Malthus on Charles Darwin. In *An Essay on the Principles of Population* (1798), Malthus had argued that human populations tend to grow exponentially and are therefore naturally subject to check by famine and poverty. Darwin generalizes this principle in positing a 'struggle for existence' in which all living things are engaged. Darwin describes his debt to Malthus as follows:

> Hence, as more individuals are produced than can possibly survive, there must in every case be a struggle for existence, either one individual with another of the same species, or with the individuals of a distinct species, or with the physical conditions of life. It is the doctrine of Malthus applied with manifold force to the whole animal and vegetable kingdoms; for in this case there can be no artificial increase of food, and no prudential restraint from marriage. Although some species may be now increasing, more or less rapidly, in numbers, all cannot do so, for the world would not hold them.[1]

Although Malthus was making a negative claim about the limits of social reform, Darwin is making a positive claim about what drives evolution. In particular, he is claiming that inter- and intra-species competition, scarcity of resources and brute physical circumstances create the pressured conditions in which some heritable traits prove themselves over others. Without such a struggle, there would be less differentiation between fitter and less fit and thus less pressure toward increased levels of adaptation. Darwin's adaptation of Malthus provides a good example of disciplinary cross-fertilization: An idea from one field sheds light on a conceptual problem in another.

We have become leery of such exchanges, however, at least when they occur at the interface of biological and social theory. This attitude owes largely to the danger of a slippage from the descriptive to the prescriptive. Darwin himself may not have construed the struggle for existence as a mandate for colonialism or the dismantling of social services, but the conflation of population analysis and political platform is tempting to make. Here the social-biological dialectic comes full circle – from Malthus to Darwin to social Darwinism – having gained an illicit normative valence in the process. Other dangers lurk here as well, including attempts at the hostile takeover of one discipline by another.

If we do not take interdisciplinary exchange seriously, however, we may be apt to ignore or mischaracterize important aspects of scientific culture. A dismissive attitude could even lead to the anachronistic pigeonholing of evolutionists whose work provides a refreshing alternative to familiar discourses. William James is one such evolutionist. The present chapter therefore examines James's social evolutionism, demonstrating its differences from several of its more familiar conceptual neighbours. It is shown that James's social evolutionism differs in important ways from social Darwinism, sociobiology and the theory of memes. Distinguishing James's view from these positions clarifies his actual motivation, which is to posit nonlinear feedback between individuals and their social and natural environments. In fact, he is positing this nonlinear dynamic at multiple levels of analysis, where a society may be construed as a higher-level developing individual. James's understanding of evolving populations can still serve as a resource for evolutionary theory, especially as a corrective to a typically reductionistic neo-Darwinism.

What James is not saying

In 1880, his thinking suffused with evolutionary and physiological influences, James prepared a talk for Harvard's Natural History Society titled 'The Functions of Great Men in Social Evolution'. Although James seems not to have delivered the talk due to illness, he revised it into one of his earliest signed scholarly publications, 'Great Men, Great Thoughts, and the Environment' (1880).[2] This piece applies Darwinian concepts to the problem of explaining history or societal change. Indeed, the very first sentence announces a 'remarkable parallel' between social and zoological evolution.[3] Namely, James holds that 'the relation of the visible environment to the great man is in the main exactly what it is to the "variation" in the Darwinian philosophy. It chiefly adopts or rejects, preserves or destroys, in short *selects* him'.[4] This is the crux of James's position: The relationship between the 'great man' and society is like the relationship between the variation and environment on Darwin's theory. James's analogy is potentially confusing, however, when read backward through the lens of intervening theories. It is thus useful to begin by explaining what James is not claiming, in order to avoid assimilating his position to one or another familiar -ism or -ology.

Social Darwinism

If 'social Darwinism' were just a label for any social theory inspired by Darwinism, then James would clearly be a social Darwinist. James's view is inspired by Darwin, as he immediately points out. However, the term has carried a more specific meaning since being given currency in the 1940s by historian Richard Hofstadter.[5] Namely, it is a pejorative label for the view that a competitive society is justified by the fact that competition is only natural. This view may be pitched at an individual level to promote competition within a society, or at a national level to promote the subjugation of one people by another. In either case, the inference is the same: Supposing that nature is good, and supposing that nature is a struggle for existence, a social order that promotes struggle is a good one. This is a gross example of deriving an *ought* from an *is*, which contains a questionable interpretation of the *is*. The living world contains cooperation as well as competition, and neither is the exclusive 'rule' of the natural order. One thus interprets nature selectively in order to derive the ethics that one wanted in the first place.

James could be suspected of social Darwinism in this sense. He portrayed himself as a rugged American individualist, appealing to individual initiative and decrying constraints on liberties. This includes the essay under consideration, which according to James, is designed to stimulate individual energies.[6] Such comments could be construed as supporting a libertarian-cum-social-Darwinist viewpoint in which all value and responsibility is located in the individual. Also of concern is the fact that in his youth James was influenced by an author who – thanks largely to Hofstadter – would come to be viewed in the twentieth century as the most infamous social Darwinist of all: Herbert Spencer, the influential English evolutionist whose psychology was outlined in Chapter 1.

It will be instructive to begin with Spencer, as James himself tends to clarify his own positions vis-à-vis Spencer's. Unlike the humble naturalist Darwin, Spencer was a systematic philosopher whose work flowed from a grand singular vision of an integrated process of universal evolution. This universal process subsumes physical, organic and social evolution as aspects of itself. It is also teleological in that it leads inexorably from homogeneity to heterogeneity. Spencer's vision of increasing heterogeneity is represented by the economic metaphor of the *division of labour*, or the idea of achieving optimum results through the specialization of parts within a whole. In Spencer's *The Principles of Psychology* (1855), this means an increase in cognitive correspondence through the association of ideas that map relations in the external world. In his *Data of Ethics* (1879), it means the development of society toward a utopia of freely acting and yet harmoniously interacting individuals. Given this grand vision, Spencer is one of the nineteenth-century's greatest advocates of a metaphysically inflated notion of progress. In Spencer's utopia, altruism and egoism have converged and violence has disappeared. In the first edition of his radical early work *Social Statics* (1851), Spencer even argues that this will entail the communal ownership of all land.

This is far from what we normally think of as social Darwinism. Nevertheless, Spencer did posit a harsh path to a gentle utopia. Evolution for Spencer is heading in a particular direction, and to promote traits that slow down the mutual adaptation of individuals is to stand in the way of progress. Spencer did not go as far as to oppose private charity or to advocate for the slaughter of the weak, but he did oppose the English Poor Laws based on the idea that such supportive services encourage individuals to pass on their shiftlessness, aggressiveness or other undesirable attributes to future generations. Given Spencer's belief in the inheritance of acquired characteristics, undesirable traits could even become entrenched in previously well-adjusted individuals and then passed on to their offspring. This makes dangerous cultural forces a potent and quickly acting influence on the fate of the species.

What James admired in Spencer was not his opposition to state services but his doggedly evolutionary vision. In James's words, 'Spencer was the first to see in evolution an absolutely universal principle.'[7] No prior writer had developed the idea with such patience and industry. Spencer brought the organism down to earth by tethering it historically and conceptually to its environment, even if – as James criticized – he portrayed causal influence in this relationship as a one-sided affair. This was a radical move, which in James's words gave 'the study of the mind in isolation a definitive quietus'.[8] Spencer's *Principles of Psychology* introduced the modern ecological sense of the term 'environment' to the English-speaking world.[9] Indeed, James's use of 'environment' in his title 'Great Men, Great Thoughts, and the Environment' must at the time have come across specifically as a reference to Spencer's term of art. Spencer also did much to popularize the term (and idea of) 'evolution', making descent with modification safer for Darwin and others. Indeed, calling Spencer a 'social Darwinist' is unfair because it downplays his own evolutionary programme, which predates Darwin's *Origin* and was primarily Lamarckian rather selectionist.[10] Social Darwinism might just as well have been called social Spencerism, although better yet is not to name such a broad concept for any concrete individual whose views are bound to outstrip it.

Spencer's strength was also his weakness, however. His monomaniacal evolutionism ultimately makes him a deductive and procrustean thinker. Spencer's ethico-political vision is part of a grand system that is inspired primarily by his theoretical predilections rather than evidence. James thus portrays Spencer as an incorrigible metaphysician who claims everywhere to find evidence but really is just incapable of seeing disconfirmation.[11] In contrast to Spencer's hyper-Victorian faith in progress, James's evolutionism is radically open-ended and posits no absolute teleology. It also entails no substantive ethical conclusions but instead suggests an analysis of meaning and value as contingent historical products. James thus rebukes Spencer by claiming that

> different ideals, instead of entering upon the scene armed with a warrant ... appear only as so many brute affirmations left to fight it out upon the chess-board among themselves. They are, at best, postulates, each of which must depend on the general consensus of experience as a whole to bear out its validity. The formula which proves to have the most massive destiny will be the true one. But this is a point which can only be solved *ambulando*, and not by any *a priori* definition.[12]

James may believe that the best ideals emerge through a certain struggle. This is altogether different, however, from claiming that *struggle as such* should be adopted as an ideal. The struggle among ideals – which could just as well be described in non-violent terms, as a kind of sifting – may end up favouring policies very different from Spencer's. If an ethics or politics is to be derived from this conception, it is something like an open marketplace of ideals, or a democratic pluralism in which multiple perspectives are given a fair hearing. James was therefore not a social Darwinist in the familiar sense.

Sociobiology

If James's social evolutionism is not social Darwinism, then perhaps it is a kind of proto-sociobiology. Spearheaded in the 1970s by Harvard biologist Edward O. Wilson, sociobiology is defined as the interdisciplinary scientific effort to provide 'the systematic study of the biological basis of all social behavior'.[13] More specifically, sociobiology is a brand of neo-Darwinism that utilizes formal tools such as evolutionary optimization models and game theory to represent social behaviours as one type of trait comprising an animal's phenotype.

Sociobiology has garnered the most attention when applied to humans. Sociobiology raises several distinct concerns in this area. One is the danger of a slippage from the descriptive to the prescriptive that was alluded to above. The social Darwinist typically derives an *ought* from an *is* synchronically, justifying present right by present might. In contrast, the sociobiologist may commit this fallacy diachronically, using past functions to justify present behaviours. In both cases, the danger is that the putative state of nature described may not be so much an observed reality as a fantasy that stacks the deck in favour of extant reactionary norms. Critics have brought special attention, for instance, to sociobiology's potential naturalization and thus vindication of sexism and violence. A second concern is that sociobiology may deny socio-cultural systems any independent properties or dynamics that are worthy of study in their own right. This tendency is manifest in Wilson's attempts at achieving a 'consilience' of knowledge that eliminates non-biological approaches to human society and culture. As Wilson puts it, '[i]t may not be too much to say that sociology and the other social sciences, as well as the humanities, are the last branches of biology waiting to be included in the Modern Synthesis.'[14] Finally, one may extend critiques of neo-Darwinism in general to sociobiology in particular. If one adaptationist hypothesis falls out of favour, is it simply replaced by another one, ad infinitum? Are researchers ignoring non-selectionist factors, such as epigenetics or developmental constraints?[15]

The point of critiquing sociobiology is not to deny painful truths. Rather, as Philip Kitcher has urged, '[t]he dispute about human sociobiology is a dispute about evidence'.[16] Given sufficient evidence, we should accept what we learn about ourselves, on pain of the most fatuous wilful ignorance. Everything turns, however, on what constitutes evidence. What kinds of factors are relevant to explaining social behaviour, and how, if at all, can one weight their relative importance? This is no trivial set of concerns, and there is no consensus about how to address them. To gain any traction here requires going beyond our twin 'nature–nurture' caricatures: first, the genetic determinist who posits a one-to-one mapping from gene to trait and gives the environment no role in development; and second, the 'soft' humanist or social scientist for whom mind or body is a blank slate that is infinitely rewritable by culture. No one holds either of these senseless views when pressed. For their part, biologists hold that the genotype codes for a *norm of reaction*, or a range of phenotypes across different environments. However, there may be fundamental problems with the very notion of a genetic code, especially if 'information' is treated in a preformationist fashion or given special causal primacy among other factors.[17]

It is possible to connect James to sociobiology through an abbreviated history of modern behavioural science. James was an influential psychologist who subscribed to Darwin's theory of natural selection (as outlined in Chapter 1). He also joined contemporaries like Conwy Lloyd Morgan, James Mark Baldwin and William McDougall in positing instincts for a variety of human behaviours around the turn of the twentieth century. Indeed, James posited human instincts liberally, claiming that humans have more instincts than other animals rather than fewer. In the middle of the twentieth century, Darwinian studies of behaviour declined sharply in the US. In part this was a reaction against the simplistic and racist eugenics programmes of prior decades, including the quasi-biological ideology of the National Socialists. However, it was also because the behaviourists who dominated mid-century psychology left little place for evolved dispositions on their dogmatically conditioning-based model of learning.[18] Thus, the emergence of an explicitly neo-Darwinian programme of sociobiology represents a return to a Darwinian approach to behavioural science that was seeded by James's *Principles of Psychology* before being stunted shortly after his death by upstart behaviourists like John Watson.

Several factors complicate this story, however. First, James operates with a deflated concept of instinct, defined as an apparently purposive behaviour that does not have to be learned. Instincts for James need not be universal in a species and are mutable like any other habits after their initial appearance.[19] James also has no specific programme of correlating instinctual behaviours with particular genes or germinal materials. James's 'instincts' have little to do with sociobiology in any event. The early instinct theory of James and his contemporaries was reconceptualised in later Darwinian work on behaviour, including the ethological studies of Konrad Lorenz and Nikolaas Tinbergen – Continental figures working during the mid-century American moratorium on Darwinian behavioural science. During this period the very distinction between innate and acquired behavioural responses came under fire, most famously in Daniel Lehrman's 1953 review of Lorenz. As Lehrman argues, 'instinctive' behaviours depend upon specific environmental conditions, and there is no coherent way to tease apart the internal (innate) and external (acquired) factors in development.[20] In the wake of this deconstruction of the instinct concept, the sociobiology of the late twentieth century aligned its focus with that of neo-Darwinism more broadly. This means that questions of embryology or ontogeny were largely ignored in favour of questions of adaptive value in evolution. Sociobiology thus views behaviours principally as optimal solutions to environmental problems.

Few researches today self-apply the label 'sociobiology'. Instead, we now have sociobiology's successor fields of evolutionary psychology and human behavioural ecology. Evolutionary psychology employs a modularist conception of cognition to explain different types of behaviours as adaptations to primeval environments. Here mind consists of relatively independent modules that can be explained as separate adaptations to particular conditions.[21] Human behavioural ecology relies heavily on kinship selection theory and cost–benefit analysis to explain human behavioural variation as a result of individuals maximizing their inclusive fitness. Sociobiology thus involves a conceptual emphasis and formal apparatus that is foreign to James's project. Its successor fields also have little in common with James, except insofar as particular scientists may emphasize Jamesian points such as the construction of environments by individuals. In general, James's distinctive interest in employing evolutionary explanations is in complicating linear models and in emphasizing the potential impact of outliers that are smoothed over in any study of averages.

Even more to the point, these neo-Darwinian programmes represent *literally* biological approaches to social behaviour. James may have contributed somewhat to this field, but this is simply not the point of his social evolutionism in the 1880 essay in question. Instead, James's social evolutionism is avowedly *analogical*. Thus, when James claims that there is a 'remarkable parallel' between social and zoological evolution, he means just this – that there is something similar about these processes. This is decidedly not the same as claiming that one process is reducible to, or an example of, the other.

This brings us closer to the actual point of James's social evolutionism.

Meme theory

Another theory that examines socio-cultural change through a Darwinian analogy is memetics, or the theory of memes. Although the term 'meme' has now come to refer to certain amusing images distributed on the Internet, it was introduced into the philosophy of science by Richard Dawkins in 1976 in *The Selfish Gene*. Just as Dawkins explains biological evolution in terms of natural selection upon genes, he explains cultural change in terms of an analogous process operating on units of culture, or memes. Dawkins's examples of memes include 'tunes, ideas, catch-phrases, clothes, fashions, ways of making pots or of building arches'.[22] Memes for Dawkins are analogous to genes, and our ability to be parasitized by them is a historical product of our genetic evolution. That is, once natural selection had endowed humans with the ability to imitate one another, this cognitive infrastructure began supporting an analogous system of cultural selection. In memetic selection, the struggle for existence consists of a competition, not for food or other material resources, but for attention and space in individuals' brains. Memetic selection is thus distinct from natural selection, and it need not improve the fitness of the individuals or populations through which it works. Memes are selfish in their own right and may even decrease the Darwinian fitness of the individuals they parasitize.

Dawkins himself notes some challenges to meme theory.[23] First, it is not clear how to individuate a meme. Memes seem to combine or merge, exhibiting something like the blending inheritance embraced by nineteenth-century biology rather than the Mendelian particulate inheritance of today. Are two similar ideas really two memes or one? Memes also do not have alleles, that is, they do not compete with specific other memes for space at a particular locus in the way that genes sometimes do. Going beyond these disanalogies, a more worrisome problem is that culture may have ineliminable system-level qualities that are not captured by a reductive memetic analysis. Just as a reductionist genetic perspective might have to be supplemented by more holistic analyses of ontogeny and phylogeny, a reductive memetic analysis might have to be similarly supplemented by certain other cultural theories. More radically, one could ask why memes should be part of the story at all. What if meme theory is an explanatory scheme foisted upon the social sciences by biologists and philosophers of science with an overweening presumption in favour of Darwinian logics, rather than a good guess motivated by observed patterns in the socio-cultural world? Dawkins has left these questions for other researchers to address.[24] All in all, the idea of social or cultural evolution is not widely accepted today, at least in those areas of the social sciences and humanities that take culture as their declared object of study.[25]

Of the views considered here, meme theory is the closest to James's social evolutionism. Unlike social Darwinism, meme theory does not derive a normative position from evolutionary principles. Unlike sociobiology, it does not imply that the study of social behaviour should be brought literally under the rubric of Darwinian biology. Rather, like James's view, meme theory is founded upon an analogy: *Socio-cultural change is like Darwinian evolution.*

James and Dawkins are not analogizing from the same concept of evolution, however. In particular, they do not share a view about what is selected in natural selection. Central to Dawkins's position is the idea that both genes and memes are *replicators*, that is, entities capable of having copies of themselves made with high fidelity across many generations. Replicators owe their success to their *vehicles*, which are entities that interact with the environment such that their affiliated replicators replicate differentially.[26] The paradigmatic replicators are (evolutionarily defined) genes, whereas the paradigmatic vehicles are the organisms that they build to inhabit. Dawkins argues that the gene is a uniquely qualified unit of selection in biology precisely because only it among biological entities is a replicator. Unlike, for example, organisms or groups, which are sadly ephemeral, genes (or copies thereof) can live on indefinitely. Dawkins gives memes this honour in the cultural domain, just as he gives genes this honour in the biological domain. Many theorists follow Dawkins in holding that evolutionary explanations must posit some entity whose replication is the measuring stick of evolutionary change.[27] One virtue of replicators is that they provide a simple and intelligible way of tracking the state of the population in question: One simply measures the relative frequency of the various types of replicators and notes how these numbers change over time.

James does not share Dawkins's background in twentieth-century population genetics or his notion of a potentially immortal self-replicating unit. If anything, Darwinism for James stands as a scientific vindication of viewing the world as an ineluctable flux containing no permanent forms. When James discusses 'variations', he is not referring to genes or germinal materials but rather to individuals and overt phenotypic traits. That is, he is using the term in a nineteenth-century sense to refer to countable things ('a variation'), not in the neo-Darwinian sense of variation as a statistical measure of genetic difference within a population. James's view thus squares better with a 'recipe' view of evolution than a replicator view: As long as certain ingredients are present – for instance, heritable variation and differential selection – then a system may be described as evolving.[28]

James therefore does not posit replicators in his social evolutionism. In particular, it would be a mistake to think that James is construing 'great men' (or any individuals) as the replicators of social evolution.[29] James's analogy must mean something different entirely.

What James is saying

The point of James's social evolutionism is not to prescribe struggle as an ideal (social Darwinism); to reduce the social to the biological (sociobiology); or to explain cultural change in terms of self-replicating units (meme theory). It is to critique the explanatory programme of Canadian writer Charles Grant Blairfindie Allen ('Grant Allen'), a follower of Herbert Spencer. Allen explains 'national characters' such as that of the ancient Greeks in geographical terms, and he gives no role to the individual in producing changes within a given society.[30] In opposition to this view, James argues that '[s]ocieties of men are just like individuals, in that both at any given moment offer ambiguous potentialities of development'.[31] Similarly, individuals *within* a society are relatively independent of the social circumstances that this society provides. There is looseness in the system, at multiple levels.

James agrees that individuals are largely a product of their social and material circumstances, but he also allows a role for individual novelty. He tends to make his point by emphasizing what he calls 'geniuses' or 'great men' who move history forward. This way of speaking calls to mind two James family friends: Thomas Carlyle, whose histories are emblematic of the Great Man theory that organizes history around the triumphs of powerful male figures; and Emerson, whose 'Uses of Great Men' (1850) exalts important figures but also celebrates the potentials of all individuals. James is channelling each of these thinkers in his celebration of individuality and in his attempt to retain some vestige of the Great Man theory. The latter theory may now strike us as simplistic, outmoded and reactionary – not to mention sexist. James's point, however, is only that individuals make some difference, such that history may be explained at least partially in terms of their particular acts.[32] Such a view was already panned as unscientific in the nineteenth century by the likes of Spencer and Allen. To reverse the charge, however, it is not necessarily scientific to make large-scale unverifiable hypotheses about national character while claiming blandly that 'everything is a result of environment'. Explanations in terms of individuals' actions at least point to concrete particulars. They are in this sense empirical. The Spencerian position, in contrast, is 'a metaphysical creed', which reverts 'to a pre-darwinian type of thought'.[33]

James's theory of social evolution is thus closely allied to his Darwinian psychology. In each case James defends the autogenous nature of individual variation. Variation brings something new to the table that the environment can select but not produce. This is selection in James's deflationary Darwinian sense. The environment's 'function is simply selective'.[34] Environments may eliminate characteristics, but they cannot produce them. This is true of a physical environment or a social one. As a result, the society *needs* input from the individual sources of variation in order to remain supple and dynamic. James thus claims that social evolution is the

> resultant of two wholly distinct factors – the individual, deriving his peculiar gifts from the play of physiological and infra-social forces, but bearing all the power of initiative and origination in his hands; and, second, the social environment, with its power of adopting or rejecting both him and his gifts.[35]

Another way to put this point is that James is telling the social philosopher not to explain too much qua social philosopher: 'He must simply accept geniuses as data, just as Darwin accepts his spontaneous variations'.[36] Every science must take certain data as basic. Darwin's great conceptual move was to expand the domain of what the environment *does not* explain by positing non-directed variation. This refusal of explaining too much with a single tool set Darwin apart from the Lamarckians, and it is a lesson to those who fall too much in love with a single mode of explanation or direction of causation.

It is interesting to note that James is not analogizing the individual person in social evolution to the individual organism in Darwin's theory. Rather, James is analogizing society to the organism and the individual person to the variation. In other words, *individuals in society are like variations in a developing organism.* James makes this clear in a diagram accompanying his essay. This diagram shows society as an individual whose development is inflected by non-directed variations, which are themselves individual human beings (whose own development is inflected by non-directed variations).[37] This is an explicitly hierarchical view, where selections at one level become variation at another. The sources of all of this variation are ultimately physiological and inscrutable.

Social evolution is then the development of a social organism among other social organisms (societies or nations) within a heterogeneous physical environment. It might seem awkward from today's perspective to call this 'social evolution', given that this 'evolution' is modelled on ontogeny rather than phylogeny.

This makes sense assuming nineteenth-century usage, however, as the terms 'evolution' and 'development' were used interchangeably at this time (such that an embryo might be said to *evolve* or a species to *develop*). The reading of social evolution as a society's ontogeny also gels with James's tendency to use Darwinian concepts principally to say something about individuals rather than populations. James is interested in non-directed variation that emerges at any point in ontogeny, regardless of phylogenetic consequences. Even where such variation is not congenital or genetically heritable, it may impact other evolving systems including society or human knowledge. Variation always may count for *something*, in some system that embeds the individual. Biology in the strictest sense is not always the point.

This essay leaves James with a view that is not only hierarchical but also dialectical. Neither individual nor environment is fully independent. The individual thus alters the environment 'just as the advent of a new zoölogical species changes the faunal and floral equilibrium of the region in which it appears'.[38] To borrow a metaphor from Richard Levins and Richard Lewontin, the environment is not like a rigid lock that individuals are shaped to fit like a key.[39] Whether in the biological or social case, individuals change the shape of the lock that they are trying to fit, in an endless dynamic feedback process.

Although James does not deduce social Darwinist norms from evolution, he does believe that his social evolutionism has a certain hortatory upshot. Namely, as mentioned above, he thinks that it 'forms an appeal of the most stimulating sort to the energy of the individual'.[40] By this James means that his view will tend to embolden individuals to be more efficacious than they might otherwise be. This is a characteristically Jamesian point that runs throughout his writings and at least once raised him from suicidal despair: To posit one's own agency is the first step in exercising this agency.[41] James's social evolutionism is thus not only an appeal for a more sophisticated conception of societal change; it is also an attack on writers whose simplistic accounts are damaging because morally enervating.

Conclusion

The question of the proper range of evolutionary concepts is a contentious one. Initiatives at the interface of biology and social theory continue to raise hackles. James's social evolutionism provides an alternative to more familiar discourses, however. James does not follow social Darwinism in gleaning brutal prescriptions from evolutionary theory. Such a view implies that we are in danger of not being natural and thus must make efforts to be more like (a certain image of) nature. Ironically, this view participates in a kind of essentialism that is deconstructed by a properly Darwinian approach. Neither a species nor nature has an ideal normative essence against which one may be measured as deficient. Species are provisional forms that emerge from natural processes that include both competition and cooperation. If one views nature as 'essentially' competitive or cooperative, this says more about one's predilections than it does about nature. James is also neither a reductionist sociobiologist nor a meme theorist. The former is a literal attempt to explain social behaviours in Darwinian terms rather than a view based on a Darwinian analogy, and both views presume a neo-Darwinian framework that James did not hold.

The point of James's social evolutionism is that societies are like developing organisms in that they are subject to non-directed variations; and these variations are themselves individuals that are subject to their non-directed variations. The point is to complicate a linear worldview by making its moving parts relatively independent while also dialectically entwined. This opens up a logical space for individual-level explanations. If James emphasizes the changes wrought by the individual, this is to balance the views of others who over-emphasize the environment. The times make the individual, but the individual also makes the times. James is especially keen to argue that major figures in history have made a difference that would not have been made without them. If Shakespeare had been stillborn, history would not have delivered a similar figure with a similar literary output simply because the times were conspiring to produce this. There was no 'Shakespeare node' waiting to be filled, just as in biological evolution there are no 'species nodes' but only a nonlinear branching process in which speciation events occur through unique and particular circumstances.

It should be clear at this point that James was not a social Darwinist in the traditional sense of the term. Nevertheless, it will be instructive to consider one more definition of this term in concluding the present chapter. In his study *Social Darwinism in American and European Thought* (1997), historian Mike Hawkins defines social Darwinism as a broad metaphysical outlook characterized by the idea that temporality and development are pervasive and significant features of the world. One need not subscribe to literal Darwinism in order to be a social Darwinist in this sense, although Darwin – alongside Hegel, Spencer and others – did help to set its conceptual foundations. This version of social Darwinism is neutral among political positions, as it may equally underwrite sanguine optimism about progress or a reactionary fear of society's imminent degeneration.

It is important to see that James *is* a social Darwinist in this sense.[42] In fact, the entire present study can be viewed as an examination of James's social Darwinism – so to speak – insofar as it examines James's construal of mind, society, reality and truth as complex dynamic systems operating in time. It is a central feature of James's pragmatism, pluralism and radical empiricism that the world is *in the making* and must be understood as such.

Notes

1 Darwin 1998/1859, 91.
2 WB, 335. This essay can be confusing to track based on its title. It appeared as 'Great Men, Great Thoughts, and the Environment' in *The Atlantic Monthly* in 1880 and then in translation in *Critique Philosophique* in 1881. The title was shortened to 'Great Men and Their Environment' when it was collected in *The Will to Believe* in 1897 (WB, 163–189).

3 WB, 163.
4 WB, 170.
5 Hofstadter 1944.
6 WB, 183.
7 EPH, 114.
8 EPH, 121.
9 Pearce (2010) argues that the term 'environment' introduced a singular noun that could be paired with the singular noun 'organism'. This conceptual dyad is subtly different from relating an organism to a collection of plural 'circumstances'.
10 For Spencer's pre-Darwinian evolutionism, see the first edition of *Principles of Psychology* (1855) and Progress: Its Law and Cause (1857).
11 EPH, 110–111.
12 EPH, 17.
13 Wilson 1980, 4.
14 Wilson 1980, 4.
15 Gould and Lewontin 1979.
16 Kitcher 1985, 9. Emphasis removed.
17 Susan Oyama's critique leads to a reframing of development that makes it difficult to draw a strict line between the genetic and the environmental – including the cultural – in the first place. See Oyama (2000a; 2000b/1985).
18 For more on this narrative, see Richards (1987, Chapter 11).
19 PP II, 1004–1057.
20 Lehrman 1953.
21 Tooby and Cosmides 1990; Barkow *et al.* 1992.
22 Dawkins 1989/1976, 192.
23 Dawkins 1989/1976, 195–197.
24 Some nuance has been added by the programme of gene-culture coevolution, or dual-inheritance theory. This programme posits interaction between biological and cultural evolution and formalizes this interaction mathematically. This work is still much more theoretical than empirical, however, and it arguably reifies a nature/culture dichotomy through its distinction between social and genetic channels of influence on the phenotype. See Cavalli-Sforza and Feldman 1981; Boyd and Richerson 1985.
25 For critical voices, see Sober (1992) and Fracchia and Lewontin (1999). For proponents of socio-cultural evolution, see Aunger (2001) and Blackmore (2000).
26 'Vehicle' owes to Dawkins (1989/1976). See Hull (1980) for the slightly different concept of 'interactor'.
27 Hull 1980; Hodgson 2005; Hodgson and Knudsen 2006; Nelson 2007; Aldrich *et al.* 2008. For critical appraisals, see Godfrey-Smith (2000) and Nanay (2011).
28 Godfrey-Smith 2009. In a classic Darwinian recipe, Lewontin (1970) lists the ingredients as phenotypic variation, differential fitness and heritable fitness. If organisms vary in a manner that causes differences in their capacity to survive and reproduce – and if offspring tend to resemble parents more than they resemble the general population – then evolution by natural selection will occur, *ceteris paribus*.
29 More amenable to a replicator analysis is James's theory of truth, which posits ideas that spread socially. Such an epidemiology of knowledge is not the point of James's 1880 essay, however. See Chapter 4.
30 Allen 1878a; Allen 1878b.
31 WB, 171.
32 James emphasizes that even small individual differences are important in a follow-up essay titled 'The Importance of Individuals' (WB, 190–195). Indeed, the importance of a difference is relative to the individuals who are interested in the difference. The importance of a difference thus has no absolute measure.
33 WB, 188–189.
34 WB, 178.

35 WB, 174.
36 WB, 170.
37 WB, 437.
38 WB, 170.
39 Levins and Lewontin 1985, 98.
40 WB, 183.
41 Perry 1935 I, 323.
42 Indeed, Hawkins (1997) lists James as a social Darwinist (120).

References

Aldrich, H. E., G. M. Hodgson, D. L. Hull, T. Knudsen, J. Mokyr and V. J. Vanberg. 2008. In Defence of Generalized Darwinism. *Journal of Evolutionary Economics* 18, no. 5: 577–596.

Allen, G. 1878a. Hellas and Civilisation. *Gentleman's Magazine* 245, 156–170.

Allen, G. 1878b. Nation-Making: A Theory of National Characters. *Gentleman's Magazine* 245, 580–591.

Aunger, R., ed. 2001. *Darwinizing Culture: The Status of Memetics as a Science.* Oxford: Oxford University Press.

Barkow, J. H., L. Cosmides and J. Tooby. 1992. *The Adapted Mind: Evolutionary Psychology and the Generation of Culture.* Oxford: Oxford University Press.

Blackmore, S. J. 2000. *The Meme Machine.* New York: Oxford University Press.

Boyd, R. and P. J. Richerson. 1985. *Culture and the Evolutionary Process.* Chicago, IL: University of Chicago Press.

Cavalli-Sforza, L. L. and M. W. Feldman. 1981. *Cultural Transmission and Evolution: A Quantitative Approach.* Princeton, NJ: Princeton University Press.

Darwin, C. 1859. *On the Origin of Species By Means of Natural Selection, Or The Preservation of Favoured Races in the Struggle for Life.* London: John Murray.

Dawkins, R. 1989. *The Selfish Gene.* 2nd edn. Oxford: Oxford University Press. Original edition, 1976.

Emerson, R. W. 1850. Uses of Great Men. In *Representative Men: Seven Lectures,* 1–21. London: George Routledge & Co.

Fracchia, J. and R. C. Lewontin. 1999. Does Culture Evolve? *History and Theory* 38, no. 4: 52–78.

Godfrey-Smith, P. 2000. The Replicator in Retrospect. *Biology and Philosophy* 15, no. 3: 403–423.

Gould, S. J. and R. C. Lewontin. 1979. The Spandrels of San Marco and the Panglossian Paradigm: A Critique of the Adaptationist Programme. *Proceedings of the Royal Society of London. Series B, Biological Sciences* 205, no. 1161: 581–598.

Hawkins, M. 1997. *Social Darwinism in American and European Thought, 1860–1945: Nature as Model and Nature as Threat.* New York: Cambridge University Press.

Hodgson, G. M. 2005. Generalizing Darwinism to Social Evolution: Some Early Attempts. *Journal of Economic Issues* 39, no. 4: 899–914.

Hodgson, G. M. and T. Knudsen. 2006. The Nature and Units of Social Selection. *Journal of Evolutionary Economics* 16, no. 5: 477–489.

Hofstadter, R. 1944. *Social Darwinism in American Thought: 1860–1915.* Pittsburgh, PA: University of Pennsylvania Press.

Hull, D. L. 1980. Individuality and Selection. *Annual Review of Ecology and Systematics* 11, 311–332.

James, W. 1880. Great Men, Great Thoughts, and the Environment. *Atlantic Monthly* 46, 441–459.

James, W. 1881. Les grands hommes, les grandes pensées et le milieu. *Critique Philosophique*, January–February 1881.

James, W. 1978. *Essays in Philosophy*. The Works of William James. Edited by F. Burkhardt, F. Bowers and I. K. Skrupskelis. Cambridge, MA: Harvard University Press.

James, W. 1979. *The Will to Believe and Other Essays in Popular Philosophy*. The Works of William James. Edited by F. Burkhardt, F. Bowers and I. K. Skrupskelis. Cambridge, MA: Harvard University Press. Original edition, 1897.

James, W. 1981. *The Principles of Psychology*. 2 vols. The Works of William James. Edited by F. Burkhardt, F. Bowers and I. K. Skrupskelis. Cambridge, MA: Harvard University Press. Original edition, 1890.

Kitcher, P. 1985. *Vaulting Ambition: Sociobiology and the Quest for Human Nature*. Cambridge, MA: MIT Press.

Lehrman, D. S. 1953. A Critique of Konrad Lorenz's Theory of Instinctive Behavior. *The Quarterly Review of Biology* 28, no. 4: 337–363.

Levins, R. and R. C. Lewontin. 1985. *The Dialectal Biologist*. Cambridge, MA: Harvard University Press.

Malthus, T. R. 1888. *An Essay on the Principle of Population: Or, A View of Its Past and Present Effects on Human Happiness*. London: J. Johnson. Originally published, 1798.

Nanay, B. 2011. Replication without Replicators. *Synthese* 179, no. 3: 455–477.

Nelson, R. R. 2007. Universal Darwinism and Evolutionary Social Science. *Biology and Philosophy* 22, no. 1: 73–94.

Oyama, S. 2000a. *Evolution's Eye: A Systems View of the Biology-Culture Divide*. Science and Cultural Theory. Edited by B. H. Smith and R. E. Weintraub. Durham, NC: Duke University Press.

Oyama, S. 2000b. *The Ontogeny of Information: Developmental Systems and Evolution*. 2nd edn. Science and Cultural Theory. Edited by B. H. Smith and R. E. Weintraub. Durham, NC: Duke University Press. Original edition, 1985.

Pearce, T. 2010. From 'Circumstances' to 'Environment': Herbert Spencer and the Origins of the Idea of Organism-Environment Interaction. *Studies in History and Philosophy of Biological and Biomedical Sciences* 41, no. 3: 241–252.

Perry, R. B. 1935. *The Thought and Character of William James*. 2 vols. Boston, MA: Little, Brown, and Company.

Richards, R. J. 1987. *Darwin and the Emergence of Evolutionary Theories of Mind and Behavior*. Chicago, IL: University of Chicago Press.

Sober, E. 1992. Models of Cultural Evolution. In *Conceptual Issues in Evolutionary Biology*, ed. E. Sober, 535–551. Cambridge, MA: MIT Press.

Spencer, H. 1851. *Social Statics: Or, the Conditions Essential to Human Happiness Specified, and the First of Them Developed*. London: Chapman.

Spencer, H. 1857. Progress: Its Law and Cause. *Westminster Review* 11, 445–485.

Spencer, H. 1872. *The Principles of Psychology*. 2 vols. 2nd edn. London: Longmans. Original edition, 1855.

Spencer, H. 1879. *The Data of Ethics*. London: Williams and Norgate.

Tooby, J. and L. Cosmides. 1990. On the Universality of Human Nature and the Uniqueness of the Individual: The Role of Genetics and Adaptation. *Journal of Personality*, 58, no. 3: 17–67.

Wilson, E. O. 1980. *Sociobiology: The Abridged Edition*. Cambridge, MA: Harvard University Press.

3   Self-transformation
Habit, will and selection

In 1875 James penned a letter to Harvard President Charles Eliot in which he proposed to teach the first undergraduate physiological psychology course ever to be offered in the US. James imagined this course as part of a new philosophical anthropology, or as he put it, 'a real science of man' based on 'the theory of evolution and the facts of archaeology, the nervous system, and the senses'. James's question to Eliot was the following:

> The question is shall the students be left to the magazines, on the one hand, and to what languid attention professors educated in the exclusively literary way can pay to the subject? Or shall the College employ a man whose scientific training fits him fully to realize the force of all the natural history arguments, whilst his concomitant familiarity with writers of a more introspective kind preserves him from certain crudities of reasoning which are extremely common in men of the laboratory pure and simple?
>
> Apart from all reference to myself, it is my firm belief that the College cannot possibly have psychology taught as a living science by anyone who has not a first-hand acquaintance with the facts of nervous physiology. On the other hand, no mere physiologist can adequately realize the subtlety and difficulty of the psychologic portions of his own subject until he has tried to teach, or at least to study, psychology in its entirety. A union of the two 'disciplines' in one man, seems then the most natural thing in the world, if not the most traditional.[1]

When James claims to unite two disciplines, he is referring to his background in both psychology and physiology. By 'psychology', however, he means a relatively non-experimental study of mind that is still closely allied to – and taught in the same department as – philosophy. By combining these perspectives, James is proposing to make psychology and philosophy more physiological *while also making physiology more psychological and philosophical*. Right at the dawn of physiological psychology in the US, James saw this field not as a reducer of other approaches to mind and behaviour but as a way of thickening up a broadly naturalistic discussion about organisms and their capacities. At stake, among other things, is the possibility of a full-blooded interdisciplinary study of the human being as a concrete purposive organism.[2]

James's physiological investigations resulted in an innovative analysis of the individual. According to James, the sensorimotor 'reflex arc' comprises a hierarchical series of selectionist systems that is structured by habits and mediated by a selective will. This analysis provides the basis for James's ethics of self-transformation, which lies at the centre of his philosophical vision. The result is the outline of a viable moral philosophy with concrete consequences for pedagogy – taken both in the narrow sense of educational theory and in the broader sense in which philosophy is intended to offer a general theory of living and dying well.

The present chapter makes a case for this interpretation by tracing James's early physiological concepts through his writings on psychology, pedagogy and religion. These writings – especially *The Principles of Psychology*, *Talks to Teachers on Psychology* and *The Varieties of Religious Experience* – represent the core of James's thinking. As such, they provide necessary background for interpreting his other writings on such topics as pragmatism, belief and radical empiricism. Notably, they provide the proper context for interpreting his argument in 'The Will to Believe' that it is sometimes allowable to accept a belief in the absence of coercive evidence.

Spiritual crisis and resolution

James's philosophy is rooted deeply in his spiritual crisis of the late 1860s and early 1870s. During this period, James was increasingly imbibing of scientific materialism. This was partially a youthful revolt against his father's Swedenborgian mysticism. Henry James, Sr. was a high-minded amateur philosopher with little respect for either the mimetic fancies of artists or the crude materialism of scientists. He would watch his son proceed through both of these phases before eventually becoming something more akin to his metaphysician father as an adult. James's flirtation with materialism was also encouraged by his education in the new science of experimental psychology, which he had studied in Germany for a few months in 1867/1868 before practically founding the discipline in the US in the 1870s. He even drafted an article in 1872 – apparently his first attempt at writing a signed scholarly essay for publication – defending the consistency of the view that individuals are essentially automata within a deterministic material world.[3]

James was never temperamentally suited to materialism, however. As he once expressed to his friend Oliver Wendell Holmes, Jr., he believed a materialistic worldview to be incompatible with optimism or with a sensitive nature.[4] Despite his gregariousness and self-styled hardihood, James was a sensitive and chronically ill individual who struggled to adopt a form of optimism that never came to him as a free gift. Not surprisingly, therefore, James's most materialistic phase correlates with the worst of his suicidal depression and extreme panic fear. During this time James writes to his father, for instance, that he is occupied by 'thoughts of the pistol, the dagger and the bowl';[5] and to his friend Thomas Ward that he had recently been 'on the continual verge of suicide'.[6] Decades later, James was still so haunted by an experience from this time period that he recounted it at length in *The Varieties of Religious Experience* (1902):

> Whilst in this state of philosophic pessimism and general depression of spirits about my prospects, I went one evening into a dressing-room in the twilight to procure some article that was there; when suddenly there fell upon me without any warning, just as if it came out of the darkness, a horrible fear of my own existence. Simultaneously there arose in my mind the image of an epileptic patient whom I had seen in the asylum, a black-haired youth with greenish skin, entirely idiotic.... There was such a horror of him, and such a perception of my own merely momentary discrepancy from him, that it was as if something hitherto solid within my breast gave way entirely, and I became a mass of quivering fear.[7]

This dramatic portrait illustrates the depth of James's suffering during this time and the philosophical interpretation that he gave it. James would never reject the idea that moderate differences of degree separate the 'healthy' from 'mad', but he did come to believe that individuals have some say over the directions in which they turn.

In fact, by the time of this experience James had already begun to plot his escape from his philosophically inflected dread of existence. James's way out is represented by two key remarks that he makes in letters from this time period to Thomas Ward. The first is that education in the broadest sense consists in 'getting orderly habits of thought'.[8] The second remark – a more personal one – is James's contention that his sole motivation for living was now his belief in his ability to effect change in the world.[9] These remarks may seem to be of little significance taken in isolation, but they represent the germ of James's mature philosophy: A meaningful life consists in the efficacious wilful mediation of habits in the direction of one's ideals. These are the rudiments of James's ethics of self-transformation. What James lacked at this time was a specific conception of habit and volition. His position had an outline but was short on specific moving parts.

In a curious coincidence, James would borrow these moving parts largely from three texts first published in 1859.[10] One of these is Charles Darwin's *On the Origin of Species*. As detailed in Chapters 1 and 2, Darwin provides James with a naturalistic account of mental evolution as well as a conceptual framework for understanding mind as a selective agency in its own right. The second text of 1859 is the second essay of Charles Renouvier's *Essais de Critique Générale*. Renouvier was a neo-Kantian who defended freedom of the will and denied that reality could be accounted for by any single principle. Renouvier thus encouraged James's belief in the efficacy of conscious thought, as well as his pluralistic denial that the universe forms a totalized whole that could ever be known synoptically.

Although James first reports reading Renouvier in 1868, it was his 1870 reading of the second essay of his *Essais* that helped to deliver him from his spiritual crisis and sparked a lifelong friendship between the two men.[11] James prized Renouvier's elegantly simple doctrine that freedom consists in the mind's selectively attending to one idea among others. James treats this as a revelation in his diary:

> I think that yesterday was a crisis in my life. I finished the first part of Renouvier's second 'Essais' and see no reason why his definition of Free Will – 'the sustaining of a thought *because I choose to* when I have other thoughts' – need be the definition of an illusion. My first act of free will shall be to believe in free will.[12]

James believed this to be a form of freedom that was not susceptible to veto by materialistic science. Renouvier's position requires only that one admit the mind's ability to attend to its own ideas, thereby determining what motor consequences will follow. A determinist might claim that even the mind's intra-psychic workings are fated, but this claim would be as much an unverifiable metaphysical posit as its denial. James thus felt licensed to construe freedom as a live option.

The effect of this idea on James was profound. As his father wrote to his brother Henry in 1873, William was now 'animated' and 'restored to sanity', explicitly crediting Renouvier's doctrine of freedom.[13] In a notice of the same year in *The Nation*, James celebrates Renouvier's idea of founding a philosophy upon a free affirmation of freedom. This means that 'we have an *act* enthroned in the heart of philosophic thought'.[14] To base philosophy upon an act rather than a putative foundational truth is to supplant universal rationality with a ground-less affirmation of possibility. This denial of the self-sufficiency of the intellec-tual is one meaning of James's pragmatism.

James's auto-affirmation of freedom is familiar to many of his readers. James's diary entry also contains a less-cited passage, however, in which he elaborates on the structure of the self that is altered by the will's free acts:

> For the present then remember: care little for speculation; much for the form of my action; recollect that only when habits of order are formed can we advance to really interesting fields of action – and consequently accumulate grain on grain of wilful choice like a very miser; ... Today has furnished the exceptionally passionate initiative which Bain posits as needful for the acquisition of habits.[15]

This passage shows that, from the very outset, James's conception of freedom was about not just will but *will-in-relation-to-habit*. Habits give shape and meaning to the will's possibilities, just as habits are shaped by the will over time. They are therefore crucially important for any theory of pedagogy or moral development. This holistic picture of the reciprocal influence of will and habit is indispensable for understanding James's philosophy. As James indicates, his major source for the habit side of his view is Scottish physiological psychologist

Alexander Bain. Indeed, Bain's *The Emotions and the Will* is the third text of 1859 upon which James builds his ethics of self-transformation.

Darwin, Renouvier and Bain: These are not passing influences on James's thinking. James actually reviews both Renouvier's *Essais* and Bain's *The Emotions and the Will* in a single 1876 essay in which he argues that physiological thinkers like Bain ought to adopt Renouvier's overtly anti-determinist position.[16] This is just the synthesis of Renouvier's doctrine of freedom and Bain's theory of habit – structured by innovative extrapolations from Darwinian biology – that James would place at this centre of this thought.

Physiological psychology

James was a trained physiologist whose only earned degree was in medicine. It therefore should not be surprising that his philosophy is a physiologically grounded one. His analysis of the individual is based upon the nineteenth-century physiological idea of the reflex arc, combined with a Renouvier-style conception of will that is interpreted along Darwinian lines. In mediating a self that is comprised by habits, the will for James is both free and (in a sense) creative.

Reflex action and selectionism

According to the 'reflex action' or 'reflex arc' doctrine, the sensorimotor system is a tripartite structure where cognition functions as a middle term between sensation and behaviour. The structure comprises an arc or – as John Dewey would insist – a *circuit* in which sensation, cognition and behaviour are defined in terms of their functional interrelations.[17] Certain scholars have recognized the importance of this idea for James. Ellen Suckiel, for instance, claims that reflex action undergirds James's idea that individuals are essentially 'conative, striving, desiring, purposive, idealizing, and goal-oriented';[18] and James Pawelski argues that reflex action is an overlooked 'hermeneutic key' to James's work.[19] These assessments are correct: Reflex action – especially as combined with the generalized logic of selectionism – is absolutely central to James. It is also the clearest source of the pragmatist conception popularized by James that thinking is for action.

James pays heed to reflex action in his earliest signed essays. In 'The Sentiment of Rationality' (1879), for instance, he notes that the 'structural unit of mind is in these days, deemed to be a triad, beginning with sensible impressions, ending with motion, and having a feeling of greater or less length in the middle'.[20] In 'Reflex Action and Theism' (1881) James invokes reflex action almost to chasten cognition and put it in its place. This issues in a functional account of mind:

> The willing department of our nature, in short, dominates both the conceiving department and the feeling department; or, in plainer English, perception and thinking are only there for behavior's sake.[21]

It is difficult to overstate how radical this position is in relation to traditional philosophy: Mind is neither essentially mimetic (as in empiricism) nor speculative or transcendental (as in rationalism and idealism) but rather instrumental and practical – a function of a nervous system that has evolved to help a certain lineage of organisms get by in the world.

Reflex action is closely tied to another nineteenth-century physiological notion, which William Benjamin Carpenter dubbed 'ideo-motor action'. If the sensorimotor system comprises an arc from sensation to cognition to action, then ideo-motor action occurs when this arc is traversed without conscious intervention. The ideo-motor doctrine ascribes an inherent momentum to nervous impulses, which discharge along paths of least resistance. In the case of ideo-motor action, there is so little resistance that the entire process occurs automatically. In James's words, ideo-motor action occurs '[w]herever movement follows *unhesitatingly and immediately* the notion of it in the mind'.[22] This is not to say that mental operations are bypassed but rather that sensation passes immediately to associated mental representations that discharge immediately in associated motor consequences. That such action should be possible makes evolutionary and intuitive sense: Ideo-motor action is more efficient than continually overwhelming the nervous system with reflection upon familiar and easily navigable circumstances.

James adapts the reflex action and ideo-motor concepts to his own purposes. In particular, he structures them using another model that the present study has already examined at some length: generalized selectionism. The reflex arc for James is selectionist in structure, at multiple levels of analysis. He therefore claims that 'Selection is the very keel on which our mental ship is built'.[23] The point here is not just the widely recognized one that *selective attention* is ubiquitous in James's psychology, although this is true.[24] It is also that the selectivity of mental functioning instantiates a general model that James abstracted from Darwin's theory and posited at various levels throughout the reflex arc (and indeed throughout his worldview). In short, the Jamesian reflex arc comprises a *hierarchical selectionism*.[25]

This structure is in place at least as early as James's 1878 Lowell Lectures 'The Brain and the Mind'.[26] James restates this view in *The Principles of Psychology* (1890), where he contends that 'consciousness is at all times primarily a *selecting agency*'.[27] James thus analogizes the reflex arc to a series of filters, where selection at one level provides variation at the next. James traces this selective activity in detail, starting with sensation:

> To begin at bottom, what are our very senses themselves but organs of selection? Out of the infinite chaos of movements, of which physics teaches us the world consists, each sense-organ picks out those which fall within certain limits of velocity. To these it responds, but ignores the rest completely as if they did not exist.[28]

The sense organs are strict gatekeepers whose physico-chemical structure picks out specific features of the world.[29] This process is massively selective in that the overwhelming majority of potentially sensible features of the world go unregistered and thus count for nothing mentally.

From sensation James passes to the perception of objects as such:

> If the sensations we receive from a given organ have their causes thus picked out for us by the confirmation of the organ's termination, Attention, on the other hand, out of all the sensations yielded, picks out certain ones worthy of its notice and suppresses all the rest.[30]

From a sea of already-selected sensation, attention fashions tractable objects of perception. This further gerrymanders sensation into a collection of substantive, nameable things. The mind does not produce the sensations to which it attends but only structures and foregrounds them. In doing so, it tends to suppress what James calls the transitive 'flights' of experience (such as feelings of logical, spatial or temporal relation) in favour of the substantive 'perchings' (abstract concepts and perceived objects).[31] It is a key point of both James's psychology and his metaphysics of radical empiricism, however, that experience is sensibly continuous. Relations for James are as real as substantive objects or abstractions, even if one tends to overlook them. This is why consciousness for James is a *stream*, as opposed to the *train* of atomistic ideas posited by classical empiricism. Discrete ideas are a refined product of abstraction, ironically posited as the basic building blocks of mental life.

Next the mind determines what it will count as an object's essential features: 'The mind selects again. It chooses certain of the sensations to represent the thing most *truly*'.[32] For example, in order to perceive a circular red table as such, one must abstract from the table's heterogeneous brightness and shading, taking a specific shade of red to be its true colour; one must also take its shape viewed directly from above (or below) to be more truly its shape than the oval profile it has from other angles; etc. James holds that such essences may shift from time to time or person to person, as he views essences as practical expedients rather than absolute realities.

These initial three levels of selection – sensory, objective and essential – are fundamental. They are preconditions for perceiving a world of discrete and tractable objects among which overt action may take place. James also mentions three more levels of mental selection that may occur once this perceived world is given. First, reasoning in general

> depends on the ability of the mind to break up the totality of the phenomenon reasoned about, into parts, and to pick out from among these the particular one which, in our given emergency, may lead to the proper conclusion.[33]

Second, in artistic practice, 'The artist notoriously selects his items, rejecting all tones, colors, shapes, which do not harmonize', such that the resultant artwork is mainly 'due to *elimination*'.[34] Finally, there is the ethical selection of one's character or self:

Ascending still higher, we reach the plane of Ethics, where choice reigns notoriously supreme. An act has no ethical quality whatever unless it be chosen out of several all equally possible. To sustain the arguments for the good course and keep them ever before us, to stifle our longing for more flowery ways, to keep the foot unflinchingly on the arduous path, these are characteristic ethical energies.[35]

This passage captures the outline of James's moral philosophy: The moral life consists in selecting among competing interests that represent different possible ways of being.

Notably, the ordering of these levels of selection is logical or analytical rather than chronological. One does not first sense and then think and then act, before repeating this sequence again from the start. All levels are operating at all times and influencing each other in a nonlinear fashion. For example, even if sensation is prior to attention in the scheme of the reflex arc, habits of attention nevertheless influence what gets sensed and how sensations are categorized. The nonlinearity of these processes demonstrates the limits of James's filtration metaphor. According to James,

The highest and most elaborated mental products are filtered from the data chosen by the faculty next beneath, out of the mass offered by the faculty below that, which mass in turn was sifted from a still larger amount of yet simpler material, and so on.[36]

This metaphor captures the non-directedness of variation, as a filter is defined by its function of selecting (rather than producing) elements with particular properties. However, a properly functioning filter is typically understood to be static rather than malleable. The metaphor therefore breaks down because the 'filters' of the nervous system are indeed altered by the types and intensities of the elements passing through them. Nervous channels do not merely direct the flow of impulses; the impulses also reshape the channels. This is not a flaw in the nervous system. On the contrary, it is what allows for plasticity and growth. Indeed, it takes being bombarded repeatedly with associated sensations for the mind to learn to recognize different kinds of objects in the first place, lest it be caught forever in the 'blooming, buzzing confusion' that James attributes to the newborn infant.[37]

If the nonlinear quality of reflex action weakens James's filter analogy, it actually strengthens the analogy with natural selection. Although natural selection is distinguished from Lamarckian instructionist explanations in that it does not rely on the environment to induce adaptive variation directly, the environment in natural selection does affect variation indirectly. Specifically, it does this by entrenching developmental constraints that allow only a certain range of variation to become manifest in the organism and thus to become 'visible' to natural selection. Something similar happens in the Jamesian reflex arc, where mental variation at lower levels may be directed or constrained by the prior selective activity of higher levels. Indeed, this is just what allows for self-transformation, or the top-down biasing of future possibilities by the individual.[38]

The point in both the biological and psychological cases is not to deny any kind of environmental influence on the individual. It is to recognize this influence as loose and indirect. This is why positing selection is always a *complexifying* move for James, as against an attitude of causal austerity that steamrolls the messy details of complex systems.

Will

James's conception of will in the *Principles* is an elaboration of Renouvier's view mentioned above. For both Renouvier and James, willing means attending to an idea, especially where multiple competing ideas are present. This means preserving an idea and thus favouring its associated motor consequences:

> The essential achievement of the will, in short, when it is most 'voluntary', is to ATTEND to a difficult object and hold it fast before the mind. The so-doing is the *fiat*; and it is a mere physiological incident that when the object is thus attended to, immediate motor consequences should ensue.... Effort of attention is thus the essential phenomenon of the will.[39]

The sensorimotor system mostly runs automatically, but the will weighs in where deliberation is required. In other words, the will intervenes in cases of obstructed ideo-motor action. One could imagine a bottleneck of ideas-cum-actions that the will unclogs. James prefers a different metaphor, however, which is that of tending a flame:

> The idea to be consented to must be kept from flickering and going out. It must be held steadily before the mind until it *fills* the mind. Such filling of the mind by an idea, with its congruous associates, *is* consent to the idea and to the fact which the idea represents. If the idea be that, or include that, of a bodily movement of our own, then we call the consent thus laboriously gained a motor volition. For Nature here 'backs' us instantaneously.[40]

The end-result of this process may either be overt action or the cancelling of an action. After all, the inhibition of one impulse by another is a genuine motor consequence that requires real nervous energy. James himself was a chronic vacillator who knew that apparent inaction might actually be the result of complex and enervating deliberation.

Willing for James is an explicitly moral matter. In fact, it is the very fulcrum of one's moral agency. 'Moral' for James is not principally a label for certain types of action, but rather a name for applying volition in the service of some ideal. On James's view, '*To sustain a representation, to think,* is, in short, the only moral act'.[41] This means that an admirable act, done by rote habit, is not moral on James's view (which is not to call it *immoral*). On the other hand, any sincere victory in self-overcoming contains all of the drama and significance of the moral life. As explored in Chapter 4, James joins Nietzsche as a philosopher of striving who finds morality not in pat formulas but in the cultivation of a changing self in relation to a changing world.

The Jamesian will does not achieve such effects by virtue of a supernatural power. On the contrary, James explicitly denies the existence of a substantive soul, transcendental ego or other unexplained explainer. The will for James is an evolved function, the chief role of which is to stabilize and intelligently direct a complex organism. It is characteristic of complex systems that small changes in conditions can produce large-scale unpredictable effects. James argues that the human nervous system in particular is unstable and erratic. This is a positive feature in that it makes individuals rich in non-directed mental variation that feeds creativity and intelligence. However, it also necessitates the will's function of weighting certain possibilities over others. James thus claims that the will 'loads the dice' in favour of interests posited by consciousness: '*Can consciousness increase its efficiency by loading its dice? Such is the problem*'.[42] Armed with a will, the individual benefits both from wide possibilities and the ability to narrow these possibilities for chosen purposes.[43] The will is in this sense a 'fighter for ends' within an unstable nervous system.[44]

James puts forth this view in opposition to the epiphenomenalist 'conscious automaton theory'. According to Thomas Huxley, for instance, consciousness is to the nervous system as steam is to a train: a by-product that does not reach back down to influence the machinery that created it. As indicated above, James himself formulated such a theory and defended its consistency in his first attempt at drafting a scholarly article for publication in 1872. By the time of his 1875 review of Wilhelm Wundt, however, James had articulated his Darwinian objection to this view: 'Taking a purely naturalistic view of the matter, it seems reasonable to suppose that, unless consciousness served some useful purpose, it would not have been superadded to life'.[45] James echoes this position in his 1879 essay 'Are We Automata?'[46] and in the *Principles*, where the epiphenomenalist position is associated with both Huxley and William Clifford. As odd as it may sound today, James invokes Darwinism to defy mechanistic reductionism, rather than viewing this metaphysics as part and parcel of modern science.

Today we would call James's Darwinian functionalist account of consciousness an *adaptationist* hypothesis, or speculation about a trait's origin based on its apparent function. Adaptationist methodology has justly received scrutiny within the philosophy of biology,[47] and James himself generally treats such hypotheses with caution. In any event, the merit of James's hypothesis depends upon the specifics of the case. One question is whether consciousness might plausibly constitute a non-selected corollary of other traits, or a structural feature that has become entrenched for reasons other than its apparent use. If so it could plausibly have tagged along while lacking efficacy, or it might not be meaningfully individuated as a 'trait' in the first place. On the other hand, it is arguable that the constellation of processes in which volition is embedded is too complex and integrated for it to be a mere evolutionary side effect. James's account also gels with our intuitive sense that conscious decisions are qualitatively different from habitual or ideo-motor action. If volition is something additional to non-volitional activity – and if the ability to choose deliberately with foresight increases the ability to survive and reproduce on average – then there might be something to James's argument.

Habit

James does not put forth this Darwinian functionalist model of volition merely in order to account for piecemeal decision-making. He is also interested in how the will's selections alter the very self in which it operates. James makes it clear that his emphasis is on the self-constructive activities of the individual. When the individual makes a choice about a specific matter, the deeper choice 'really lies between one of several equally possible future Characters. What he shall *become* is fixed by the conduct of this moment'.[48] In short, 'The problem with the man is less what act he shall now choose to do, than what being he shall now resolve to become'.[49] James is thus concerned with the *existential* dimension of willing, or the fact that a decision never leaves the decider unchanged. Less sophisticated accounts of rational action may ignore this complexity, just as less sophisticated versions of Darwinism ignore the activities of organisms in constructing their environments.

What constitutes the self that is being changed? The answer for James is *habit*. James defines habit in the *Principles* with Carpenter's simple statement that 'our nervous system grows to the modes in which it has been exercised'.[50] This growth consists in the strengthening or weakening of nervous channels through repeated or diminished use. Such differential weighting of pathways occurs in multiple ways. In the case of 'instincts', a pathway may predictably become entrenched early in development (given ordinary environmental conditions). James claims that an instinct – defined as an apparently goal-oriented behaviour that was never learned – is strictly an instinct only upon its first exercise, at which point it can be considered a plastic habit among others. The initial appearance of non-instinctual habits depends more upon the specific details of the individual's life history. They are less predictable in how and whether they will arise, although they, too, are plastic and mutable after they arise.

Habit is a double-edged sword. On the one hand, it makes easier the repetition of what has come before. By constraining possibilities, it narrows one's field of vision and thereby increases efficiency. The batter in baseball swings more accurately when the relevant sequence of muscular contractions fires off in an unconsciously associated chain; overt thinking is destructive here. Indeed, to make action ideo-motor one need only engrain a habit to a point of automaticity. Again, it is generally best for familiar situations to be handled without conscious attention. On the other hand, the conservative nature of habit also means that one constantly finds oneself stuck in what mappers of evolutionary space call a 'local optimum' – an adaptive state that has proved good enough, or better than readily available alternatives, but which could be surpassed if a broader view of the landscape could be attained. We all know certain habits that would benefit us but that are just far enough from our current constitution that they are difficult to attain. This same point can be made about individual habits and social norms, which are entwined and mutually reinforcing. James illustrates this point using the mechanical metaphor of a flywheel:

> Habit is thus the enormous fly-wheel of society, its most precious conservative agent. It alone is what keeps us all within the bounds of ordinance, and saves the children of fortune from the envious uprisings of the poor. It alone prevents the hardest and most repulsive walks of life from being deserted by those brought up to tread therein.[51]

The fact that we actually *can* resist habit and press forward toward some ideal underscores a crucial difference between wilful selection and natural selection: The former, but not the latter, can see past the immediately local in order to pursue ideals that are only vaguely glimpsed. Blind phylogeny has produced a system that is capable of some measure of foresight in ontogeny.

Freedom and creativity

As James himself points out, nothing in his Darwinian account of consciousness entails free will. The heart performs the evolved function of pumping blood, but this does not make it 'free' in any special sense. Similarly, the will might perform the evolved function of selecting upon mental variation without thereby being *free* in any significant sense. We certainly feel that we choose, but – to speak with Huxley – this might just be the feeling of the steam escaping from the whistle. James's account of the will thus leaves open the metaphysical question of freedom. It does, however, give this question a precise formulation: 'Are the duration and intensity of this effort fixed functions of the object, or are they not?'[52] We are free only if the amount of attention given to an idea is (at least in some cases) indeterminate in advance. If so, the individual makes a real difference by selecting ideas and thus actions.

James does not believe this issue to be theoretically decidable. Thus, as in his early reading of Renouvier, he treats freedom as a practical posit. This calls to mind Kant, who posits freedom while denying that it can be theoretically proven. Unlike Kant, however, James does not distinguish between a deterministic phenomenal realm and an unknowable noumenal realm. Instead, James posits that the will's selections constitute original choices within a single world that is natively loose or 'indeterministic'. There is just one world, and it is genuinely unfinished.

If James defines the will's freedom in selectionist terms, he invokes this same logic to deny the will's *creativity*. According to James, 'The soul *presents* nothing to herself; *creates* nothing; is at the mercy of the material forces for all *possibilities*; but amongst these possibilities she *selects*'.[53] This is again James's deflationary sense of 'selection': To select is merely to select, not to produce. If freedom consists in the will's curating of possibilities that it could not elicit or produce, then the will is in a certain strict sense not a creator. Its freedom is an uncreative freedom, or so it would seem.

This is a misleading statement, however, as it conflates a technical sense of 'creative' with a more colloquial one. Creativity in the normal sense does not require *ex nihilo* novelty. Such a requirement would be absurd, rendering creativity arbitrary. Creativity consists not in the conjuring of novelty from nowhere but in the *mediation* of the novelty with which one is constantly presented. That is, creative activity in the arts, sciences or other areas of human endeavour means constructing significance by weaving the novel into the familiar. Indeed, novelty is comprehensible only where it serves some such bridging function. In James's words, 'neither the old nor the new, by itself, is interesting: the absolutely old is insipid; the absolutely new makes no appeal at all'.[54] Creativity means bringing novelty to bear on extant structures of meaning and resistance. The Jamesian will is nothing if not such a mediator. It is in this sense both creative and free.

As Robert Doyle argues, James's model of the will provides an alternative to the traditional dilemma of libertarian free will and determinism.[55] On the one hand, libertarianism buys a superlative version of freedom at the expense of making choice appear arbitrary. On the other, determinism precludes freedom unless one can abide compatibilism, or the view that the predetermination of all events is compatible with freedom after all. As a third alternative, James's model allows action to be *determined* by the will but not *predetermined* at the level of the generation of possibilities. The will chooses among possibilities that it finds in its purview. It could not have delivered these possibilities to itself, and they might have been otherwise. Such a view is interesting to revisit in the context of non-deterministic interpretations of quantum mechanics, which James did not live to see. Here the problem of free will may turn out to be that of saving some kind of determination in the face of chance. This is quite the about-face, as philosophers have long presumed a classical Newtonian framework in which the problem is to salvage freedom in a world of presumed determinism.[56]

Pedagogy *sub specie boni*

James is interested in will and habit because of their practical significance. This includes their significance for self-transformation and for education. James's reflections on pedagogy can be traced to his early book reviews. James was so taken by William Benjamin Carpenter's *Principles of Mental Physiology*, for instance, that he penned three distinct reviews of this book in 1874, claiming that 'as giving a rationale of education, or the formation of mental and moral character' it is 'one of the most valuable pedagogic publications of modern times'.[57] Carpenter thus rivals Bain in James's mind as a great physiological moralist. James also discusses pedagogy in physiological terms in his 'Habit' chapter in *The Principles of Psychology*. The culmination of James's pedagogical thinking, however, is an 1892 series of lectures to educators that he published in 1899 as *Talks to Teachers on Psychology* (traditionally bound together with James's *Talks to Students on Some of Life's Ideals*). *Talks to Teachers* is an application of the primary concepts of the *Principles* – will, habit, reflex action and the stream of consciousness – to practical questions of education.

In *Talks to Teachers* James defines education itself as 'the organization of acquired habits of conduct and tendencies to behaviour'.[58] In other words, the aim of education is the same as that of moral development in general: to '*make the nervous system our ally instead of our enemy*'.[59] In this context, the teacher must provide information while also inculcating character. In James's words,

> Your task is to build up a *character* in your pupils; and a character, as I have often said, consists in an organized set of habits of reaction. Now of what do such habits of reaction themselves consist? They consist of tendencies to act characteristically when certain ideas possess us, to refrain characteristically when possessed by other ideas.
>
> Our volitional habits depend, then, first, on what the stock of ideas is which we have; and second, on the habitual coupling of the several ideas with action or inaction respectively.[60]

Teaching is thus directed by complementary aims: The first is to *solidify* students by engraining a mass of interrelated habits of thinking and behaving; the second is to make them existentially *suppler* by increasing their factual and moral imagination. Students are to be firmly planted in developmental space, while at the same time being equipped to search beyond their station for distant outposts. Sound education thus fosters precisely the kind of character that is both the means and end of James's ethics of self-transformation: firm yet malleable; stable yet imaginative.

James clearly links his pedagogical theory to his moral psychology. This is especially clear in his explication of a series of maxims that he first introduced in the *Principles* (paraphrased here).[61] James adopts the first two of these maxims from Bain's *The Emotions and the Will*, while the remaining three are his own additions:

1   Use as strong an initiative as possible when intentionally acquiring a habit.
2   Never suffer an exception to the habit until the habit is securely rooted.
3   Seize every opportunity to act on the maxim correlated with the desired habit.
4   Do not preach or talk in the abstract. Just lie in wait for practical opportunities to act.
5   Keep your faculty of attention alive by gratuitous exercise, including self-denial in unnecessary things.

Each of these maxims is meant to aid in the efficient alteration of one's habits, and each demands a certain maturity or hardihood. The fifth maxim stands out, however, because it applies not to any given habit but rather to the maintenance of the self-transformative machinery more generally. This maxim advises the intentional cultivation of conscious attention – which is precisely to cultivate *willing* in James's sense – so that this muscle is toned and available when it is needed. The key point is that one's moral capacities depend upon how one maintains them. Morality does not proceed from a universal rational faculty, or from our access to something outside of time or space. It is nothing apart from moral agency, which is everywhere a function of a concrete developing organism.

James's pedagogical theory focuses on the production of habits and thus behaviour. Nevertheless, it should not be conflated with the austere behaviourism that overtook psychology on the heels of James's death. Behaviourism explains behaviour as a result of conditioning. In doing so, it is basically externalist or 'outside-in' in its explanatory structure. In contrast, James's basic psychological programme is to limit externalism in order to make room for internalism and constructionism – that is, the autogenous and world-constructing activities of the individual. James thus emphasizes spontaneous interests and idiosyncrasies, as well as the ability to choose freely among competing interests. For James, a student can and should be moulded by external cues, which should be provided by teachers among others. However, two students receiving the same conditioning are not expected to turn out the same, even if they are conditioned identically from birth. Individuals are different from each other in ways that cannot be directly or systematically controlled by external influences. Education for James thus requires more than rule-based conditioning. It also requires a kind of intuitive work that is difficult to formalize. James thus concludes *Talks to Teachers* with the following portrait of the pupil-as-fellow-organism:

> I cannot but think that to apperceive your pupil as a little sensitive, impulsive, associative, and reactive organism, partly fated and partly free, will lead to a better intelligence of all his ways. Understand him, then, as such a subtle little piece of machinery. And if, in addition, you can also see him *sub specie boni*, and love him as well, you will be in the best possible position for becoming perfect teachers.[62]

Teachers must help students refine their ability to mediate their own unruly mental variation. Doing this effectively involves setting a positive example and empathizing with them in a holistic fashion. The best teachers are thus socially and emotionally intelligent, as well as technically adept.

James's use of the phrase *sub specie boni* ('under the aspect of the good') is interesting in this context. This phrase is a callback to a preceding reference he makes to an observation by Spinoza:

> Spinoza long ago wrote in his Ethics that anything that a man can avoid under the notion that it is bad he may also avoid under the notion that something else is good. He who habitually acts *sub specie mali*, under the negative notion, the notion of the bad, is called a slave by Spinoza. To him who

acts habitually under the notion of good he gives the name of freeman. See to it now, I beg you, that you make freemen of your pupils by habituating them to act, whenever possible, under the notion of a good.[63]

One may quit smoking to avoid the pain of illness, or one may do so out of a positive vision of flourishing health. The distinction is between shrinking from what is hated and growing toward what is desired. To adopt the latter orientation is to choose *sub specie boni*. James's recommendation is to adopt the less defensive and more hopeful attitude and to encourage students to do the same.

James's pedagogical theory and his ethics of self-transformation are of a piece, and both build upon the moral psychology of *The Principles of Psychology*. The major difference is that education is principally other-directed (by teachers), whereas self-transformation is essentially self-directed. Ideally, the former process sets a useful habitual structure for the latter process to build upon. However, self-transformation is sometimes required to reconstruct habits acquired through a less-than-useful education. If education grafts a second nature onto the first, then the moral autonomy developed in adulthood might deliver a third nature.

The will to believe

Self-transformation is also central to James's well-known essay 'The Will to Believe' (1897). The point of this much-abused piece of writing is not that one may hold a belief in the face of disconfirming evidence, or still less that one may force one's unconfirmed beliefs on others. On the contrary, James's point is that there are highly circumscribed cases where it would be absurd to veto an individual's compulsion to believe if this would render impossible certain avenues of living that this belief opens up. In contrast to his adversary William Clifford, who argues that belief on insufficient evidence is in all cases wrong, James likens belief to a *bet* that one is free to undertake at one's own risk. As in other areas of life, an overly risk-averse attitude is just as foolish as an overly risky one. There is no duty to avoid error at all costs, since one must also factor in the potential benefits of accepting a belief that is uncertain. 'The Will to Believe' thus provides a further example of James's emphasis on the benefits of acting *sub specie boni* as opposed to shrinking from negative possibilities *sub specie mali*.

James was long interested in the capacity of beliefs to make realities possible. In his 1878 essay 'Quelques considérations sur la method subjective', for example, he adduces the case of the Alpine hiker whose voluntary belief in his ability to cross a chasm makes that very jump possible.[64] In the 'Will to Believe', James is specifically interested in what he calls a 'genuine option', defined as a decision that is *living* in that the individual is genuinely capable of accepting different answers; *forced* in that it is not logically avoidable (as in a strict disjunction); and *momentous* in that it is unique in one's life and has irreversible practical consequences. James argues that in the case of a genuine option it is permissible to hold a belief in the absence of confirming logical or empirical evidence. Individuals are free, for example, to posit a divine presence and reap the benefits of the strenuous life. James makes a similar point in *Some Problems of Philosophy* (1911), arguing that any philosophy that vetoes belief in a genuinely improvable ('melioristic') world is absurd for the practical reason that *if such a world existed it would bar us from believing this.* James concludes, 'Faith thus remains as one of the inalienable birthrights of our mind'.[65]

'The Will to Believe' is not an isolated contribution to a niche ethical debate but an elaboration on James's earlier moral psychology. In fact, it is not too much to claim, as Colin Koopman does, that 'The Will to Believe' and key chapters in *The Principles of Psychology* are actually studies of the self-same process of self-transformation under different descriptions.[66] The concept of 'belief' in 'The Will to Believe' is belief as *lived.* Just as Peirce investigates the relation between the irritation of doubt and the 'resting place' of belief,[67] James is interested in the concrete process in which an idea is attended to, deliberated upon and accepted. Belief is not a passive state of accepting something to be true, as in an abstract list of things that one would assent to if asked. Nor is 'will' in this essay interchangeable with other terms like 'resolve', 'drive' or 'wish'. As James had already lain out in the book that made his career, the self is structured by a reflex arc that is governed by a selective will. On James's view, to select an idea and thus its motor consequences requires attending to it wilfully. Thus, will in 'The Will to Believe' should be understood as the evolved function of will to which James had devoted detailed descriptions in the *Principles*. After all, a Jamesian genuine option requires wilful selection by definition, since it cannot discharge in an automatic or ideo-motor fashion.

'The Will to Believe' is thus a *case study* in James's ethics of self-transformation, which examines a particular category of decision-making. This category has special characteristics, to be sure. The decision to accept (say) a spiritual interpretation of the universe is not as simple as the decision to reach forward to pick up a glass of water. It is more likely to be the result of extended or interrupted deliberation, and its motor effects may consist in a substantially reconfigured habitual structure instead of a single overt action. The highly philosophical and self-reflexive character of this case does not exempt it from being an example of wilful selection, however. On the contrary, it is a particularly important example of the latter. James's ethics of self-transformation gives special weight to decisions that significantly reconstruct one's character, as opposed to smaller and more piecemeal choices. In 'The Will to Believe', as elsewhere, James is interested in decisions insofar as they reconfigure the self that will go on to make future decisions.

'The Will to Believe' nicely complements another of James's essays from the 1890s, 'On a Certain Blindness in Human Beings'.[68] This essay's titular blindness refers to individuals' systematic inability to appreciate what is profound or significant in the lives of others. James recommends countering this blindness in both a negative and positive sense. That is, he claims that his essay absolutely forbids us to be forward in pronouncing on the meaninglessness of forms of existence other than our own; and it commands us to tolerate, respect, and indulge those whom we see harmlessly interested and happy in their own ways, however unintelligible they may be to us.[69]

Countering blindness in individuals and society is the essence of both pluralistic democracy and pluralistic philosophy. Indeed, one of James's deepest commitments is that both knowledge and reality are pluralistic 'all the way down'. This means that certain aspects of reality may be vouchsafed only to some – and perhaps only in certain moments – while appearing obscure, uninteresting or absurd to others. Individuals therefore ought to show each other a modicum of respect regarding the unproven beliefs that make a difference in their lives. In James's words,

> Hands off: neither the whole of truth, nor the whole of good, is revealed to any single observer, although each observer gains a partial superiority of insight from the peculiar position in which he stands. Even prisons and sickrooms have their special revelations.[70]

James emphasizes that scientists in particular should not be treated as general-purpose experts on knowledge. Although a trained physiologist himself, James has no patience for chauvinistic scientism. He therefore bristles against thinkers like Huxley who take an arrogant or dismissive tone in wielding their cultural authority. In an 1874 notice in *The Nation*, for instance, James pokes fun at the idea of 'physiologist Blank, who, having got tired for a time of the laboratory's confinement, now appears in his new and brilliant role of Blank, the Audacious and Ingenious Speculative philosopher'.[71] Decades later, James retains this attitude in a letter to his friend Charles Strong:

> When you defer to what you suppose a certain authority in scientists … I am surprised. Of all insufficient authorities as to the total nature of reality, give me the 'scientists', from [Hugo] Münsterberg up, or down. Their interests are most incomplete and their professional bigotry immense. I know no narrower sect or club, in spite of their excellent authority in the lines of fact they have explored, and their splendid achievement there. Their only authority *at large* is for *method* – and the pragmatic method completes and enlarges them there.[72]

The willingness to accept the results of empirical inquiry has been a huge advance for society, insofar as this has been achieved. To understand the scientific method in terms of the pragmatic method, however, means contextualizing scientific inquiry within the greater span of human concern (as outlined in Chapter 5). In this context, technical expertise is only sometimes of special use. Indeed, a great virtue of science according to James is precisely its democratizing stance that 'no man is an expert, no man an authority'.[73] If no one is an authority *within science*, then being an accomplished scientist certainly does not grant one authority on matters of broader interest.

The Varieties of Religious Experience

Originally two series of Gifford Lectures given by James at Edinburgh University in 1901/1902, *The Varieties of Religious Experience* is an influential study of religion as the individual experiences it. In James's words, the *Varieties* is about 'the feelings, acts, and experiences of individual men in their solitude, so far as they apprehend themselves to stand in relation to whatever they may consider the divine'.[74] This topic turns out to be very broad, as James defines 'the divine' generously enough to include Emersonian transcendentalism, atheistic Buddhism and more. Religion for James means one's total way of responding to the universe, or one's particular feeling of being at home in the world. As suggested by the book's subtitle 'A Study in Human Nature', the *Varieties* is not a study of religious institutions or dogmas per se but a work of philosophical anthropology.

Religion is interesting to James because it reflects highly general features of individual experience. James picked up this idea from another psychologist and religious theorist, Edwin Starbuck, whose *Psychology of Religion* he had written a preface for in 1899.[75] By performing statistical analysis upon survey results – a method that James resisted due to its tendency to 'average out' individual differences – Starbuck had demonstrated correspondences between religious conversions and experiences of normal adolescent development. In fact, religious experiences may serve the function of catalysing and scripting personal crises that typically occur at this stage of life in any event. If one is going to have a crisis, it may as well occur in a culturally sanctioned fashion that offers a pre-given system of meaning.

James thus follows Starbuck in embedding his examination of religion in a more general examination of normal processes of psychological development. He claims that in adolescence, for instance, the 'evolution of character' consists in

> the straightening out and unifying of the inner self. The higher and lower feelings, the useful and erring impulses, begin by being a comparative chaos within us – they must end by forming a stable set of functions in right subordination.[76]

The self tends toward fragmentation during adolescence but becomes realigned in some new configuration as one reaches adulthood. Something similar may happen in religious conversions, as well as in other crises such as an intense romantic entanglement, political awakening or recovery from addiction. In any of these processes, deconstruction is required for reconstruction to occur. One does not reach a new stable configuration without dismantling an existing one, and the discomfort that attends such dismantling is not inherently an evil. Reconstruction is ongoing, and pain is not pathology.[77]

James's talk of conversion experiences in the *Varieties* adds complexity to his account of self-transformation. Here James moves beyond particular habits to groups of habits that he calls 'centres of energy'. A centre of energy for James is a relatively stable set of inter-associated cognitive-affective-behavioural habits that define an outlook or mode of being. Individuals have multiple centres of energy, which may 'take stage' in different circumstances. No one behaves the same in all circumstances or all social contexts. The *habitual* centre of energy is distinguished, however, by being the dominant or default mode:

> It makes a great difference to a man whether one set of his ideas, or another, be the centre of his energy; and it makes a great difference, as regards any set of ideas which he may possess, whether they become central or remain peripheral in him.[78]

In other words, one's habitual centre of energy defines one's *character*. One's character defines one's entire existence, including, according to James, the basic premises of one's philosophy. This is the meaning of James's claim in *A Pluralistic Universe* that philosophical interpretation means catching the author's 'centre of vision'.[79] In a very general way, one's centre of energy generates what James calls one's *Binnenleben* ('underlying life'): 'This inner personal tone is what we can't communicate or describe articulately to others; but the wraith and ghost of it, so to speak, are often what our friends and intimates feel as our most characteristic quality'.[80]

In this context, self-transformation does not just mean wilfully mediating particular habits. It also means understanding habits to be clustered into relatively stable and self-perpetuating sub-selves that represent entire worldviews or modes of being. One typically does not resist an entrenched habit by itself but rather the entire sub-self to which it belongs. This shows why it is not enough to resist a particular habit *sub specie mali*. One must also promote an underdog sub-self *sub specie boni* so as to promote it within the hierarchy. Otherwise, there is not enough leverage to displace the existing structure. In a classic case of religious conversion, for example, one may reject sensualism in favour of an austere commitment to ideal ends. A new key is established, and the dissonant notes are either muted or bent into shape.

The process of establishing a new habitual centre of energy should be understood in terms of another framework that James borrows from Starbuck. This is the distinction between conscious (or volitional) and unconscious (or self-surrender) types of conversion.[81] If volitional conversion is a conscious and goal-oriented process, then self-surrender conversion means allowing an unconscious process to resolve. Volitional conversion is easier to assimilate to James's earlier moral psychology. Volitional conversions are those that one can bring about intentionally through the will's capacity for self-transformative activity. In the *Varieties*, however, James emphasizes at least two limitations of such activity. First, attempts at volitional conversion too often proceed myopically and *sub specie mali*. One hates one's current state but has no clear picture of how to proceed. Second, challenging one's most entrenched habits is difficult *by definition* and therefore tends to occur only irregularly and reluctantly. These factors together tend to make volitional self-transformation disoriented and scattershot. In contrast, self-surrender conversion is guided by the momentum of unconscious tensions toward a crisis. It is like the realignment of tectonic plates accommodating internal pressures and strains. It is going where it needs to go whether the individual knows it or not. Narrowly egoistic volition may even obstruct its progress: 'So long as the egoistic worry of the sick soul guards the door, the expansive confidence of the soul of faith gains no presence'.[82] James even claims that a stage of self-surrender is necessary for conversion in all cases, including those that are principally volitional in earlier stages. At some point, one must learn to get out of one's own way.

James thus gives pride of place to self-surrender conversion. In doing so, he displays a certain faith in the holistic integrity of the organism. As Kurt Goldstein would argue in the 1930s, the organism displays a remarkable capacity for adjusting to injury and disease.[83] This capacity goes beyond mere homeostasis, or the maintaining of an extant equilibrium. For James, as for Goldstein, it also comprises a self-actualizing or growth-oriented dimension, which includes the ability to repurpose extant structures for new ends.[84] James's trust in the unconscious and the bodily contrasts with the pessimism of his contemporary Freud, and it prefigures the humanistic psychology movement of the middle of the twentieth century. A healthy alignment of embodied values is one dimension of the overall health toward which the organism moves. To obstruct this movement is literally to make oneself ill.

Conclusion

James builds his philosophy of self-transformation upon a physiological foundation. The first plank is Darwin's theory of natural selection. Natural selection allows James to ascribe to consciousness an evolved function and gives him a logic that he generalizes throughout the reflex arc. The second plank is Renouvier's doctrine that volition consists in attending to an idea so as to allow it to discharge in its associated motor consequences. This plank dovetails with the first one in two ways: Phylogenetically speaking, the will is a physiological function which has evolved by natural selection; and ontogenetically speaking, it achieves this purpose by selecting in a manner that is analogous to natural selection. The third plank is Bain and Carpenter's conception of the self as a plastic bundle of habits. Self-transformation in this context means utilizing selective willing to mediate one's own habitual structure. One learns how to do this in part from one's teachers and other role models, who should inculcate the habit of willing in the service of some ideal (*sub specie boni*) rather than negatively or blindly. Ideally, however, one develops some measure of independence from one's own education and acculturation. One may exert this autonomy, for instance, in 'willing to believe' a view that is empirically undecidable but which opens up new avenues of living.

James's psychology does not entail freedom of the will but gives it a particular formulation: To be free would be to select wilfully upon cognitive variation, where neither the content of the variation nor the amount of attention paid is predetermined. In his affirmation of freedom, therefore, James posits that variation is inflected by genuine novelty and that the will makes a real difference in making the selections that it does. The selectionist logic of Jamesian freedom may rule out creativity in a certain technical sense – the same sense in which no selective environment creates what it selects – but it does not rule out creativity in the quotidian sense. Nothing is recognized as creative that does not in some way interweave the novel into the familiar, which is just what the Jamesian will does. In this way, constraints on creativity – including the brute givenness of one's ideas and the specificity of one's historical context – may be viewed as positive and generative as well as negative and limiting. Both freedom and creativity mean *negotiating* constraints, not lacking them.

The topic of self-transformation unearths a tension in James's philosophy. James is generally considered to promote a hardy philosophy of energetic activity. Indeed, James defines moral action in the *Principles* precisely as attending to an idea when other ideas (and thus other forms of self) would come more easily. It becomes clear in the *Varieties*, however, that moral practice also means letting go so that one may be changed. Passivity thus plays a central role in James's thought. As Richard Gale has noted,[85] reconciling these modes of practice in James's philosophy is a challenging task. This question will be addressed in relation to Nietzsche in Chapter 4 and in relation to Gale in the Conclusion. Ultimately, different aspects of oneself are active and passive with regard to one another, in multiple senses of activity and passivity.

Notes

1 Quoted in Perry (1935 II, 11).
2 On James and disciplinary boundaries, see Bordogna (2008).
3 MEN, 246–256.
4 LWJ I, 82–83.
5 LWJ I, 95–96.
6 LWJ I, 129.
7 VRE, 134–135. James anonymizes this story by attributing it to a fictional 'French correspondent', but he reveals the truth in a letter to his French translator Frank Abauzit (VRE, 508).
8 LWJ I, 119–120.
9 LWJ I, 132.
10 This is not to claim that James was most influenced by the 1859 first editions of these works.
11 Renouvier 1859, §§9, 11.
12 Perry I, 147.
13 LWJ I, 169. William also credits his reading of William Wordsworth and his rejection of the idea that mental disorders must have a physical basis.
14 ECR, 266.
15 LWJ I, 148.

16 ECR, 321–326. James reviews the 1876 third edition of Bain and the 1875 second edition of Renouvier, respectively.

17 Dewey 1896. See Phillips (1971) for a history of the reflex arc concept that centres on James and Dewey.

18 Suckiel 1982, 47 and chap. 2.

19 Pawelski 2007. The present work emphasizes the reflex arc but gives pride of place to an alternative model: selectionism. Selectionism structures the Jamesian reflex arc, in addition to other systems. Pawelski makes reference to a selection model but does not give it logical pride of place in this fashion.

20 EPH, 64.

21 WB, 92.

22 PP II, 1130. James was a reader of Carpenter, but he also credits Renouvier and Hermann Lotze for introducing him to ideo-motor action. Stock and Stock (2004) depict James's *Principles* as the place where distinct British and German traditions of thinking about ideo-motor action converge.

23 PP I, 640.

24 Gale (1999), for example, claims that the 'leitmotiv' of James's philosophy is his view 'that the essence of consciousness is to be selectively attentive on the basis of what is interesting or important' (222–223). See also Suckiel (1982), Richards (1987), Seigfried (1990) and Crippen (2010).

25 This idea builds upon Schull (1996).

26 ML, 27–28.

27 PP I, 142.

28 PP I, 273–274.

29 The metaphor of gatekeeping is imperfect. Sensation is not a process of intake but one of *transduction* in which tissues are triggered to produce a nervous impulse. The sensation would seem to be different in kind from the impingement that triggers it. Such considerations led James's contemporary Nietzsche to the sceptical position that all sensation is 'metaphor' (*Übertragung*) or a 'crossing-over' from one domain to another (Nietzsche 1979/1873). For James and Nietzsche, see also Chapter 4 of the present study.

30 PP I, 274.

31 James first makes this argument in his 1884 essay 'On Some Omissions in Introspective Psychology' (EPS, 142–167).

32 PP I, 274–275.

33 PP I, 276. James calls 'sagacity' the ability to anatomize a situation into useful parts for conceptual deliberation (PP II, 957).

34 PP I, 276.

35 PP I, 276.

36 PP I, 277.

37 PP I, 462.

38 Not all levels need be equally malleable or equally subject to control by other levels. If a species evolves a new selectionist system that selects upon variation produced by an evolutionarily older one, then the newer system may constrain the activities of the older one in order to maintain the integrity of the organism. This makes the older system less independent than it had been in the evolutionary past. See Godfrey-Smith (2009, 100–103); Buss (1987).

39 PP II, 1167. Emphasis removed.

40 PP II, 1169.

41 PP II, 1170.

42 PP I, 143.

43 James also claims that wilful selection allows us to posit a shorter timespan for evolution. This is because such intelligent selection has accelerated socio-cultural evolution, allowing us to get to our current state faster than we otherwise might have (ML, 29).

44 PP I, 144.
45 ECR, 302.
46 EPS, 38–61.
47 Gould and Lewontin 1979.
48 PP I, 276.
49 PP I, 276.
50 PP I, 117. Emphasis removed.
51 PP I, 125.
52 PP II, 1175.
53 PP II, 1186.
54 TT, 70.
55 Doyle 2010.
56 Stapp 2007.
57 ECR, 274–275.
58 TT, 27. Emphasis removed.
59 PP I, 126.
60 TT, 108.
61 TT, 49–54.
62 TT, 114.
63 TT, 113.
64 EPH, 24; translation on page 332.
65 SPP, 113.
66 Koopman (forthcoming).
67 Peirce 1877.
68 TT, 132–149.
69 TT, 149.
70 TT, 149.
71 ECR, 116.
72 LWJ II, 270.
73 ECR, 116.
74 VRE, 34.
75 ERM, 103.
76 VRE, 142.
77 Polish psychologist Kazimierz Dabrowski argued this point strongly in his theory of positive disintegration, on which the reconstruction of the individual at a new level requires what we colloquially call 'growing pains'. See Dabrowski *et al.* (1970).
78 VRE, 162.
79 PU, 44.
80 TT, 119.
81 VRE, 170.
82 VRE, 175.
83 Goldstein 1934.
84 See Taylor 1984; 2002.
85 Gale 1999.

References

Bain, A. 1859. *The Emotions and the Will.* London: Parker.

Bordogna, F. 2008. *William James at the Boundaries: Philosophy, Science, and the Geography of Knowledge.* Chicago, IL: University Of Chicago Press.

Buss, L. W. 1987. *The Evolution of Individuality.* Princeton, NJ: Princeton University Press.

Carpenter, W. B. 1874. *Principles of Mental Physiology, with their Applications to the Training and Discipline of the Mind and the Study of its Morbid Conditions.* New York: D. Appleton & Co.

Crippen, M. 2010. William James on Belief: Turning Darwinism against Empiricistic Skepticism. *Transactions of the Charles S. Peirce Society* 46, no. 3: 477–502.

Dabrowski, K., A. Kawczak and M. M. Piechowski. 1970. *Mental Growth through Positive Disintegration.* London: Gryf Publications.

Darwin, C. 1859. *On the Origin of Species By Means of Natural Selection, Or The Preservation of Favoured Races in the Struggle for Life.* London: John Murray.

Dewey, J. 1896. The Reflex Arc Concept in Psychology. *Psychological Review* 3, no. 4: 357–370.

Doyle, R. 2010. Jamesian Free Will, the Two-Stage Model of William James. *William James Studies* 5, 1–28.

Gale, R. M. 1999. *The Divided Self of William James.* Cambridge: Cambridge University Press.

Godfrey-Smith, P. 2009. *Darwinian Populations and Natural Selection.* Oxford: Oxford University Press.

Goldstein, K. 1934. *Der Aufbau des Organismus: Einfuhrung in die Biologie unter besonderer Berucksichtigung der Erfahrungen am kranken Menschen.* The Hague, The Netherlands: Nijhoff.

Gould, S. J. and R. C. Lewontin. 1979. The Spandrels of San Marco and the Panglossian Paradigm: A Critique of the Adaptationist Programme. *Proceedings of the Royal Society of London. Series B, Biological Sciences* 205, no. 1161: 581–598.

James, W. 1977. *A Pluralistic Universe.* The Works of William James. Edited by F. Burkhardt, F. Bowers and I. K. Skrupskelis. Cambridge, MA: Harvard University Press. Original edition, 1908.

James, W. 1978. *Essays in Philosophy.* The Works of William James. Edited by F. Burkhardt, F. Bowers and I. K. Skrupskelis. Cambridge, MA: Harvard University Press.

James, W. 1979. *The Will to Believe and Other Essays in Popular Philosophy.* The Works of William James. Edited by F. Burkhardt, F. Bowers and I. K. Skrupskelis. Cambridge, MA: Harvard University Press. Original edition, 1897.

James, W. 1981. *The Principles of Psychology.* 2 vols. The Works of William James. Edited by F. Burkhardt, F. Bowers and I. K. Skrupskelis. Cambridge, MA: Harvard University Press. Original edition, 1890.

James, W. 1982. *Essays in Religion and Morality.* The Works of William James. Edited by F. Burkhardt, F. Bowers and I. K. Skrupskelis. Cambridge, MA: Harvard University Press.

James, W. 1983. *Essays in Psychology.* Edited by F. Burkhardt, F. Bowers and I. K. Skrupskelis. Cambridge, MA: Harvard University Press.

James, W. 1983. *Talks to Teachers on Psychology: And to Students on Some of Life's Ideals.* Edited by F. Burkhardt, F. Bowers and I. K. Skrupskelis. Cambridge, MA: Harvard University Press. Original edition, 1899.

James, W. 1985. *The Varieties of Religious Experience.* The Works of William James. Edited by F. Burkhardt, F. Bowers and I. K. Skrupskelis. Cambridge, MA: Harvard University Press. Original edition, 1902.

James, W. 1987. *Essays, Comments, and Reviews.* The Works of William James. Edited by F. Burkhardt, F. Bowers and I. K. Skrupskelis. Cambridge, MA: Harvard University Press.

James, W. 1988. *Manuscript Essays and Notes.* The Works of William James. Edited by

F. Burkhardt, F. Bowers and I. K. Skrupskelis. Cambridge, MA: Harvard University Press.

James, W. 1988. *Manuscript Lectures*. The Works of William James. Edited by F. Burkhardt, F. Bowers and I. K. Skrupskelis. Cambridge, MA: Harvard University Press.

James, W. and H. James. 1920. *The Letters of William James: Two Volumes Combined.* Boston, MA: Little, Brown, and Co.

Koopman, C. Forthcoming. The Will, the Will to Believe, & William James: An Ethics of Freedom as Self-Transformation. *Journal of the History of Philosophy*.

Nietzsche, F. W. 1979. On Truth and Lies in a Non-Moral Sense. In *Philosophy and Truth: Selections from Nietzsche's Notebooks of the Early 1870's*, 79–97. Amherst, NY: Humanity Books. Unpublished manuscript, 1873.

Pawelski, J. O. 2007. *The Dynamic Individualism of William James*. Albany, NY: State University of New York Press.

Peirce, C. S. 1877. The Fixation of Belief. *Popular Science Monthly* 12: 1-15.

Perry, R. B. 1935. *The Thought and Character of William James*. 2 vols. Boston, MA: Little, Brown, and Company.

Phillips, D. C. 1971. James, Dewey, and the Reflex Arc. *Journal of the History of Ideas* 32, no. 4: 555–568.

Renouvier, C. 1859. *Essais de Critique Générale: Deuxième Essai*. 3 vols. Paris: Au Bureau de la *Critique Philosophique*.

Richards, R. J. 1987. *Darwin and the Emergence of Evolutionary Theories of Mind and Behavior*. Chicago, IL: University of Chicago Press.

Schull, J. 1996. William James and the Broader Implications of a Multilevel Selectionism. In *Adaptive Individuals in Evolving Populations: Models and Algorithms*, eds R. K. Belew and M. Mitchell, 243–256. Boston, MA: Addison-Wesley Longman.

Seigfried, C. H. 1990. *William James's Radical Reconstruction of Philosophy*. Albany, NY: State University of New York Press.

Stapp, H. 2007. Whitehead, James, and the Ontology of Quantum Theory. *Mind and Matter* 5, no. 1: 83–109.

Starbuck, E. D. 1899. *The Psychology of Religion: An Empirical Study of the Growth of Religious Consciousness*. London: Walter Scott.

Stock, A. and C. Stock. 2004. A Short History of Ideo-Motor Action. *Psychological Research* 68, no. 2: 176–188.

Suckiel, E. K. 1982. *The Pragmatic Philosophy of William James*. Notre Dame, IN: University of Notre Dame Press.

Taylor, E. 1984. *William James on Exceptional Mental States: The 1896 Lowell lectures.* New York: Scribner.

Taylor, E. 2002. William James and Depth Psychology. *Journal of Consciousness Studies* 9, no. 10: 11–36.

4   Character ideals and evolutionary logics in James and Nietzsche

Born only two years apart, William James (1842–1910) and Friedrich Nietzsche (1844–1900) are remarkably parallel figures in the history of philosophy. Negatively speaking, both reject the empiricist view that knowledge consists in the mental reconstruction of an independent reality, the neo-Kantian transcendentalism that grounds cognition and values beyond the natural world and the scientific materialism that acts as just one more dogmatic metaphysics among others. Positively speaking, each supplants such traditional programmes with a vision of philosophy as a new kind of practical discipline that includes and enlarges upon the sciences without exempting them from critique. In doing so, both centre philosophy on the individual, construed not as passive mechanism or supernatural agent, but as multivalent, self-fashioning organism.

Nietzsche therefore provides an excellent point of comparison for further examining an ethics of self-transformation in the context of a modern scientific worldview. Such a comparison is the task of the present chapter, which examines James's and Nietzsche's respective ideals of ethical character in terms of their reactions to the evolutionary theories of their time. Both posit a kind of inner fuel for self-transformation – *energy* and *will* – and in doing so they offer similar critiques of externalist evolutionary logics. However, they draw upon different physiological models in outlining the individual's relationship to itself and to its environing world, and they promote contrasting images of the ideal individual or society. If James embeds self-transformation in a socially shared cooperative project, Nietzsche's ideal is an elite individual that negates humanity's metaphysical needs through ascetic self-overcoming. This reflects James's location of significance in the purposive mediation of ascending levels of individual and social structure for the purpose of creating a maximally inclusive world, as opposed to Nietzsche's prizing of the ennobled supra-historical individual. The two thinkers thus present starkly different options for the reconstruction of purpose and value in the wake of Darwin's deconstruction of teleology or Nietzsche's 'death of God'.

Ships in the night

Given that James and Nietzsche each hold that philosophy is a function of biography and temperament,[1] it may be worth considering certain similarities in their lives. One such similarity is that neither James nor Nietzsche received advanced training in philosophy: James was a physiologist and medical doctor who had given up his youthful aspirations to become a painter; and Nietzsche was a classical philologist whose training exposed him mainly to the ancients. Neither earned a degree in philosophy, and both learned the modern tradition through their own curiosity and initiative. Such 'amateurism' may be construed as a liability, but it encouraged both thinkers to think outside of rigid disciplinary boundaries.[2] Indeed, both James and Nietzsche call for broadly interdisciplinary studies of human capacities and values, as in the following question posed by Nietzsche in *On the Genealogy of Morals*:

> *"What light does linguistics, and especially the study of etymology, throw on the history of the evolution of moral concepts?"*
> On the other hand, it is equally necessary to engage the interest of physiologists and doctors in these problems (of the *value* of existing evaluations); it may be left to the academic philosophers to act as advocates and mediators in this matter too, after they have on the whole succeeded in the past in transforming the originally so reserved and mistrustful relations between philosophy, physiology, and medicine into the most amicable and fruitful exchange.[3]

Nietzsche thus outlines important work for both himself (the etymologist) and James (the physiologist and doctor).[4]

A second similarity is that both James and Nietzsche were 'neurasthenics' who suffered from multiple chronic ailments and severe mental crises: For instance, both were unable to use their eyes for close reading or writing for long periods of time throughout their adult lives; and while Nietzsche collapsed physically and mentally at age 44, the depressive James seems to have spent time at the McLean Asylum near Boston as a young man.[5] Not surprisingly, both thinkers construed existence as a challenge wrought with suffering, to be addressed by a philosophical rebellion against the very real threat of passivity and fatalism. Health problems also encouraged both thinkers to pursue such a rebellion in a pluralistic or perspectivist fashion. This is by their own description, as both thinkers construe illnesses as alternative modes of health that bring their own philosophical perspectives. According to James, 'No one organism can possibly yield to its owner the whole body of truth. Few of us are not in some way infirm, or even diseased; and our very infirmities help us unexpectedly'.[6] Similarly, Nietzsche reports, 'I am very conscious of the advantages that my fickle health gives me over all robust squares. A philosopher who has traversed many kinds of health, and keeps traversing them, has passed through an equal number of philosophies'.[7]

Despite their philosophical commonalities and shared historical moment, however, James and Nietzsche developed their philosophies more or less independently of one another. One piece of evidence that Nietzsche knew of James is that he seems to have encountered a comparative analysis of James and Francis

Galton in a book titled *Psychologie des grands hommes* in 1887.[8] Here author Henri Joly outlines James's disagreements with the genetic determinism of Darwin's cousin Galton and his belief in the power of idiosyncratic individuals.[9] James's individualism in this debate may have spoken to Nietzsche, as both were post-Emersonian individualists attempting to re-envision individuality through the optic of the life sciences.[10] This is speculation, however, and Nietzsche's second-hand encounter with James came near the end of his career. Perhaps Nietzsche might have known James better if he had been less averse to reading in English or had not descended into madness months before the publication of James's career-making *Principles of Psychology* (1890).

In contrast, James owned multiple of Nietzsche's mature works in the original German, in which he was fluent.[11] His correspondence also shows him developing a fascination with Nietzsche, especially toward the end of his career.[12] This fascination was encouraged by his reading of two French studies of Nietzsche,[13] and by his brother-in-law William Salter's work on one of the first such studies to appear in the US.[14] Indeed, Nietzsche shows up in James's letters increasingly in the final two years of his life, when James reports reading certain articles that made him more sympathetic to Nietzsche's views.[15] He even calls Nietzsche 'an extraordinary human being', with the proviso that '[t]he Nietzsche *cult* to me is sickening'.[16] In addition, he reports ordering more copies of Nietzsche's works,[17] and he recommends one of the French books on Nietzsche to multiple correspondents including his daughter Margaret Mary.[18] Given James's tendency to claim thinkers worldwide as part of his philosophical vanguard – Henri Bergson, F. C. S. Schiller and Giovanni Papini come to mind – his friend James Ward's question in 1909 was a reasonable one: 'By the way did it ever occur to you to claim Nietzsche?'[19]

James did not claim Nietzsche. Correspondence aside, James mentions Nietzsche only in a book review, a couple of minor essays, twice in *The Varieties of Religious Experience* and occasionally in his unpublished notes.[20] These are mainly off-handed references, and the only substantive discussion comes late in the *Varieties*. This is disappointing but perhaps not surprising. Several factors militated against a fair reading of Nietzsche in the US at the turn of the twentieth century: There were no English translations of Nietzsche's works prior to 1896, and his works were then translated out of order; early interpretation was influenced by Nietzsche's sister Elisabeth Förster-Nietzsche, an anti-Semite who lionized him while misrepresenting his views; also unhelpful was Max Nordau, a reactionary polemicist who demonized Nietzsche in a popular work of anti-modernism called *Degeneration* (*Entartung*) that was translated into English before any of Nietzsche's works were. Freshly deceased, Nietzsche was already subject to divergent and unfair caricatures.[21] Apart from defending Nietzsche from Nordau's 'abuse' in a book review,[22] James did little to rectify this situation. Indeed, scholars are only beginning to make up for the missed connection between James and Nietzsche today – an effort of which this chapter is meant to be a part.[23]

Evolutionary logics of the self

Therefore, the point of reading Nietzsche through James is not that James got Nietzsche right. It is that James's uses of Nietzsche in *The Varieties of Religious Experience* are indicative of important themes in both of their philosophies, which lead to the core of their respective views. Here it must be recognized that James and Nietzsche share a post-Darwinian ethics of self-transformation fuelled by internal powers and structured by a hierarchy of sub-selves. This means, first of all, that both James and Nietzsche construe the organism as inherently active or form-giving.

Nietzsche as 'dying rat'

Nietzsche's first function in the *Varieties* is to provide an example of an attitude that does not count as religious. Nietzsche might have appreciated being cast as 'irreligious', but not in the sense that James meant it. It must be recalled that the topic of the *Varieties* is not religious dogmas or institutions but rather the experiences of individuals 'so far as they apprehend themselves to stand in relation to whatever they may consider the divine'.[24] James defines the term 'divine' broadly enough to include the experiences of Emersonian transcendentalists and atheistic Buddhists. What interests James is individuals' most basic attitudes toward existence. Almost any serious existential attitude may be religious on James's liberal use of the term. However, Nietzsche is disqualified because of his sneering attitude:

> The mood of a Schopenhauer or a Nietzsche – and in a less degree one may sometimes say the same of our own sad Carlyle – though often an ennobling sadness, is almost as often only peevishness running away with the bit between its teeth. The sallies of the two German authors remind one, half the time, of the sick shriekings of two dying rats. They lack the purgatorial note which religious sadness gives forth.[25]

James's charge is a serious one, which strikes at the heart of Nietzsche's philosophy: Because he is flippant and scornful, Nietzsche is incapable of grappling constructively (or 'purgatorily') with the tragic dimension of existence. For this reason his attitude does not qualify as religious in James's positive sense.

It is telling that James here conflates Nietzsche with Arthur Schopenhauer. James was well acquainted with Schopenhauer's writings, having found his pessimism in turns humorous and obnoxious since he was a teenager. Schopenhauer eschewed the optimistic rationalism of other German idealists, preferring to draw upon Vedic philosophy in order to construe existence as arational suffering. The thesis of Schopenhauer's major work *Die Welt als Wille und Vorstellung* ('The World as Will and Representation') is that the phenomenal realm is an illusion suffused with pain due to its origins in a blindly striving noumenal will. The recommended escape from this primordial suffering is to quell it through some combination of ascetic self-abnegation and disinterested aesthetic contemplation. In other words, one must quiet the will within oneself.

James's attitude toward Schopenhauer is summed up in an 1883 letter in which he refuses to contribute to the construction of a Schopenhauer monument in Frankfurt:

> But is there no other man than Schopenhauer on whom we can combine? I really *must* decline to stir a finger for the glory of one who studiously lived for no other purpose than to spit upon the lives of the like of me and all those I care for.... As for Schopenhauer himself, personally, his loud-mouthed pessimism was that of a dog who would rather see the world ten times worse than it is, than lose his chance of barking at it, and whom nothing would have unsuited so completely as the removal of cause for complaint. There are pathetic pessimists and cantankerous pessimists. Schopenhauer was not pathetic.[26]

Obviously, James in the *Varieties* lumps Nietzsche in with Schopenhauer as a 'cantankerous pessimist'. In doing so, however, he misses the fact that Nietzsche's favourite reason for citing Schopenhauer was actually the same as his: to illustrate an unacceptably passive reaction to the horrors of existence. Despite his early debts to Schopenhauer – noticeable, for instance, in the Apollonian-Dionysian distinction and worship of music in *The Birth of Tragedy* (1872) – Nietzsche spent most of his career at pains to replace his elder's 'resignationist' pessimism with a joyous *Ja-sagen* that James himself echoes in the yes-saying of the saint figure in the *Varieties*.[27] This is precisely to deny James's claim that Nietzsche's philosophy lacks a purging or 'purgatorial' note. Like James, Nietzsche was engrossed by the project of outlining a meaningful mode of agency despite the prevalence of suffering in all of its forms.

Indeed, the two thinkers adopt this orientation in reaction to some of the same evolutionary debates. In particular, both are pugnacious critics of asymmetrically externalist accounts of ontogeny and phylogeny. James makes clear that this is his basic disagreement with Spencer in an 1878 letter:

> My quarrel with Spencer is not that he makes much of the environment, but that he makes *nothing* of the glaring and patent fact of subjective interests which cooperate with the environment in molding intelligence. These interests form a true spontaneity and justify the refusal of a priori schools to admit that mind was pure, passive receptivity.[28]

For James the interests that determine one's habits of attention are 'spontaneous' in that they are not explicable in terms of environmental demands. This makes perception both a biased filter for sensory input and a producer of non-directed cognitive and behavioural variation. Nietzsche echoes this criticism of Spencer's definition of life and mind as 'correspondence': 'life itself has been defined as a more and more efficient inner adaptation to external conditions (Herbert Spencer).... one overlooks the essential priority of the spontaneous, aggressive, expansive, form-giving forces'.[29]

If both James and Nietzsche hold that individuals are active and form-giving, both also associate these capacities with a fulfilling existence. According to Nietzsche, the strong are

> too wise to dissociate happiness from action ... all in sharp contrast to the 'happiness' of the weak and the oppressed, with their festering venom and malignity, among whom happiness appears essentially as a narcotic, a deadening, a quietude, a peace, a 'Sabbath', an enervation of the mind and relaxation of the limbs, – in short, a purely passive phenomenon.[30]

James sounds strikingly Nietzschean in making a similar point:

> Passive happiness is slack and insipid, and soon grows mawkish and intolerable. Some austerity and wintry negativity, some roughness, danger, stringency, and effort, some 'no! no!' must be mixed in, to produce the sense of an existence with character and texture and power.[31]

Happiness as negative relief from stress is inferior to happiness as positive exertion. Some amount of the former may be necessary to cultivate the latter, but it is no substitute for it. For a twenty-first century example, one can imagine the relatively passive and negative escape from stressors into the glow of one or another personal electronic device.

Both James and Nietzsche go as far as to claim that life is void of meaning without challenges that prompt resistance. In 'The Dilemma of Determinism' (1884), for instance, James rejects Spencer's view that society is moving inexorably to a utopia in which altruism is habitual and conflict has disappeared. In contrast, James's promotes the *melioristic* view that every advance must be fought for and won. More to the point, however, is that Spencer's gentle utopia sounds *boring*:

> The white-robed, harp-playing heaven of our sabbath-schools, and the ladylike tea-table Elysium represented in Mr. Spencer's *Data of Ethics*, as the final consummation of progress, are exactly on par in this respect – lubberlands, pure and simple.... To our crepuscular natures, born for the conflict, the Rembrandtesque moral chiaroscuro, the shifting struggle of the sunbeam in the gloom, such pictures of light upon light are vacuous and expressionless, and neither to be enjoyed nor understood.[32]

This passage is striking in the richness of its imagery. Like the mythical place of leisure in the broadside ballad 'An Invitation to Lubberland', Spencer's utopia would promote a *tedium vitae* (weariness of life) whose pure light is not suited to our crepuscular (active-at-dawn-or-twilight) or chiaroscuro (involving-the contrast-of-light) nature. James thus dramatically recommends extinction before such a fate: 'If *this* be the whole fruit of the victory, we say ... better ring down the curtain before the last act of the play'.[33] As in *The Principles of Psychology*, James is claiming that a significant life requires a certain tension. This tension results from the disconnect between an ideal and a present state, as experienced in the resistance offered by entrenched bodily and social structures. Spencer's milquetoast utopia would seem to lack the generative tension necessary for such moral meaning.

This criticism of Spencer is potentially confusing in the context of James's essay. The dilemma in James's 'Dilemma of Determinism' is not that between determinism and free will, as one might expect. It is a dilemma between two different kinds of determinism: 'pessimistic' determinism that describes existence as inherently evil and 'subjectivist' determinism that explains evil as a necessary ingredient in a good universe (in a kind of fatalistic theodicy). James prefers subjectivism to pessimism, and he makes his critique of Spencer's utopia in the voice of the subjectivist. Ultimately he rejects both kinds of determinism, however, in favour of *indeterminism,* or the view that the universe contains genuine novelty and that evil is a finite problem to be fought by free individuals in an ongoing manner. James therefore grabs both horns of the dilemma and throws it out the window. This rejection of determinism does not change James's view that moral meaning requires resistance, however. In fact, it may heighten it: Especially in an indeterministic universe, existence gets its savour from the mediation of entrenched yet plastic structures.

These comments are echoed by Nietzsche's terser remarks on Spencer just a couple of years later in the fifth book he appended to the 1887 edition of *The Gay Science.* Here Nietzsche classes Spencer among scientists who fall into a 'spiritual middle class' due to their inability to glimpse deep insights:

> Take, for example, that pedantic Englishman Herbert Spencer. What makes him 'enthuse' in his way and then leads him to draw a line of hope, a horizon of desirability – that eventual reconciliation of 'egoism and altruism' about which he raves – almost nauseates the likes of us; a human race that adopted such Spencerian perspectives as its ultimate perspectives would seem to us worthy of contempt, of annihilation![34]

Remarkably, Spencer's philosophy is enough to push both James and Nietzsche to the imagery of universal suicide, and for similar reasons. For Nietzsche, as much as for James, one purges suffering through morally meaningful individual agency.

Notably, Nietzsche immediately broadens his critique of Spencer to a general critique of mechanistic science. For Nietzsche, Spencer's positing of a 'horizon of desirability' in the convergence of altruism and egoism shows that he has not recognized the superficiality of his ethico-political theory. He compares this to how mechanistic scientists lack the 'good taste' to respect the limits of the horizon set by their methods:

Do we really want to permit existence to be degraded for us like this – reduced to a mere exercise for a calculator and an indoor diversion for mathematicians? Above all, one should not wish to divest existence of its *rich ambiguity*: that is a dictate of good taste, gentlemen, the taste of reverence for everything that lies beyond your horizon. That the only justifiable interpretation of the world should be one in which *you* are justified because one can continue to work and do research scientifically in *your* sense (you really mean, mechanistically?) – an interpretation that permits counting, calculating, weighing, seeing, and touching, and nothing more – that is a crudity and naiveté, assuming that it is not a mental illness, an idiocy.[35]

Nietzsche's defence of nature's 'rich ambiguity' (*seines vieldeutigen Charakters*) is a strong statement of what he calls 'perspectivism'. According to Nietzsche, there is no synoptic viewpoint that captures reality completely. Something always escapes. The temptation to think otherwise may be fuelled by the desire to believe that one's pet methods provide a special key to the universe. When all you have is a hammer, everything looks like a nail.

Of course, this is also a statement of James's pluralism. At a broad level, therefore, James and Nietzsche share a great deal in common in their views on both morality and knowledge – views that were forged in reaction to the same positions exemplified by Spencer.[36]

Moral fuel: energy and will-to-power

Before examining Nietzsche's second appearance in the *Varieties*, it will be useful to look more closely at James's and Nietzsche's respective analyses of the individual. Both thinkers construe the individual as a multivalent hierarchical structure deriving meaning from moral struggle. However, they posit distinct concepts of the *fuel* for moral action, and they utilize different physiological and evolutionary logics in explaining how this fuel is put to work in the world.

The fuel of moral agency on James's view is *energy*. Energy for James is a reserve of power that may be used for constructive ends, and one may have more or less of it. Although energy for James is essentially a quantitative concept, fluctuations in one's energy may cause radical qualitative shifts in one's attitudes or capacities. According to James, the most important thing that can happen to an individual is to have this level of energy raised. James thus posits a historically realized divine being in his 1891 essay 'The Moral Philosopher and the Moral Life' precisely in order to open up reserves of energy, or to foster the 'strenuous life'. The point is to marshal the resources necessary to live at one's most efficient level of moral activity, which James allows will differ from person to person.

James's clearest remarks on energy come in his 1906 address to the American Philosophical Association titled 'The Energies of Men'.[37] Citing the common experience of a 'second wind', James suggests that individuals may actually have layer after layer of energy that is sealed up due to physiological blockages or inveterate habits. He therefore recommends experimenting with 'dynamogenic' practices that cultivate energy, such as yoga, meditation and self-denial in small things. Because moral agency is a function of idiosyncratic developing organisms – not of a disembodied or universal moral faculty – it must be exercised in order to work optimally. Indeed, given that James defines moral action in *The Principles of Psychology* precisely as the resistance of sub-optimal entrenched structures, there is no morality on James's view without the energy that is expended in such resistance.

Although James considers the raising of moral energies to be a vitally important mission, he warns that dynamogenic processes are very little understood. He thus concludes his lecture with an exhortation to the systematic and interdisciplinary study of personal energies:

> The first of the two problems is *that of our powers*, the second *that of our means of unlocking them or getting at them*. We ought somehow to get a topographic survey made of the limits of human power in every conceivable direction, something like an ophthalmologist's chart of the limits of the human field of vision; and we ought then to construct a methodical inventory of the paths of access, or keys, differing with the diverse types of individual, to the different kinds of power. This would be an absolutely concrete study, to be carried on by using historical and biographical material mainly. The limits of power must be limits that have been realized in actual persons, and the various ways of unlocking the reserves of power must have been exemplified in individual lives.[38]

James is seeking nothing less than the advent of a new area of study, devoting to mapping out and exploring the nature and dynamics of individual moral efficacy. This area of study has not come to pass, except perhaps unsystematically in corners of popular or organizational psychology.

If James's moral fuel is energy, then Nietzsche's is *will-to-power*. It is tempting to read this concept in social Darwinist or proto-fascist terms, given its connotations of irrational greed or domination. However, will-to-power for Nietzsche is not a drive toward a particular kind of power (such as economic or political) but rather a general tendency toward expansion and incorporation. Will-to-power is thus not defined negatively as an emptiness that *seeks* power. It is already intrinsically power, and it seeks to grow. Nietzsche's concerns about will-to-power are more qualitative than quantitative: All agency expresses will-to-power in some way, but some forms of will-to-power seem relatively devious or perverse. In particular, Nietzsche argues that the 'slave morality' of Christianity expresses a 'weak' will-to-power. This will-to-power is not weak in the sense of being easily cowed or dominated. On the contrary, it seems to have taken over Europe. It is weak in that it convinces the spiritually exemplary to deny their instincts for enforcing certain kinds of hierarchy and in this way protects the mediocre from being subordinated. Although this vision is not fascist in a straightforward sense, it is at a deep level less democratic than James's.

Nietzsche was an avid reader of natural science, especially after his encounter with neo-Kantian Friedrich Albert Lange's *Geschichte des Materialismus* (*History of Materialism*) (1866). Nietzsche scholar Gregory Moore argues, for instance, that will-to-power is an adaptation of Anglo-German zoologist William Rolph's view that life seeks essentially to expand itself.[39] Although will-to-power is easiest to observe in the growth and procreation of organisms, Nietzsche hypothesizes that it is actually the principle undergirding all activity in nature:

> Suppose, finally, we succeeded in explaining our entire instinctive life as the development and ramification of *one* basic form of the will – namely, of will to power, as *my* proposition has it; suppose all organic functions could be traced back to this will to power and one could also find in it the solution of procreation and nourishment – it is *one* problem – then one would have gained the right to determine *all* efficient force univocally as – *will to power*.[40]

Nietzsche defends this proposal based upon parsimony: It is better to explain all efficacy in terms of a single principle rather than proliferating principles beyond necessity. Nature itself is the upsurging and ramification of will-to-power. Interestingly, this upsurging seems to pulse across generations. That is, in contrast to the neo-Darwinian view that sharply delineates ontogeny from phylogeny, Nietzsche embraces the nineteenth-century conception that procreation is actually a type of growth. Some active principle from the parents lives on continuously – as opposed to being reconstructed from scratch genetically – in offspring. This captures something of the sense in which Nietzsche was a Lamarckian, which is quite different from Spencer's externalist Lamarckism.

If will-to-power is Nietzsche's adaptation of Rolph's biology, it is also his twist on Schopenhauer's posit that the world in its noumenal aspect is will or striving. Nietzsche follows Schopenhauer in generalizing the experience of willing in order to arrive at a theory about the structure of reality. What is closest to experience is taken to explain what is farthest away. A major difference between Nietzsche and an idealist like Schopenhauer, however, is that in the final analysis Nietzsche rejects all dogmatic metaphysics. Instead, he adopts a more playful and hypothetical position. On Nietzsche's view, will-to-power is both an interpretation of reality and the source of all interpretations. Nietzsche thus claims circularly – almost gaily and mischievously circularly – that he is expressing his own will-to-power in the theory of will-to-power. Nietzsche does not claim to have a position beyond will-to-power from which to describe its functioning. Like James, he is aware that ultimate metaphysical hypotheses are improvable and are therefore justified intuitively and pragmatically.[41]

The metaphor of moral fuel does not apply as cleanly to Nietzsche as it does to James. Energy for James keeps the organic moral engine running. Nietzsche's will-to-power is an active principle inhering in the very drives that it expresses. It cannot be separated, as a container is separated from its content. Life *is* will-to-power for Nietzsche.

Moral structure: centres of energy and internal oligarchies

James and Nietzsche are both great typologists of human character. For instance, James distinguishes 'tender-minded' rationalists from 'tough-minded' empiricists in *Pragmatism*, and Nietzsche contrasts spontaneous 'masters' to reactive 'slaves' in *On the Genealogy of Morals*. Nevertheless, both emphasize that pure types do not exist because the very nature of individuality is to consist of some admixture of competing tendencies. In James's words, 'No man is homogeneous enough to be fairly treated, either for good or ill, according to the law of one "type" exclusively. There is "more" of him'.[42] The foremost ethical question for both James and Nietzsche is how to negotiate with this 'more'.

The task of James's ethics of self-transformation is to seek the best hierarchy among one's 'centres of energy', which are sets of self-reinforcing cognitive-affective-behavioural habits that represent entire sub-personalities or world-views. Adolescence marks a time when centres of energy are likely to shift, but 'conversions' can occur at any point, as in the case of a religious or romantic crisis or a recovery from addiction. As James makes clear in *The Varieties of Religious Experience*, habits are not broken or entrenched on a piecemeal basis. To engage in self-transformation requires the promoting of an underdog centre of energy that reconstructs the existing order. Otherwise there is no foothold from which to proceed.

Similarly, the stated goal of Nietzsche's *Genealogy of Morals* is to find 'the solution of the *problem of value*, the determination of the *order of rank among values*'.[43] Interestingly, whereas James tends to run the analogy of individuality upward – depicting societal change as ontogeny, for instance – Nietzsche analogizes in the opposition direction by depicting the individual as a kind of society. According to Nietzsche in the *Genealogy*, 'our organism is an oligarchy' in which it is imperative 'to make room for new things, above all for the nobler functions and functionaries, for regulation, foresight, pre-meditation'.[44] Nietzsche also describes the self (or 'me') in *Beyond Good and Evil* as a society of 'under-souls':

> *L'effet c'est moi* [The effect is me]: what happens here is what happens in every well-constructed and happy commonwealth; namely, the governing class identifies itself with the success of the commonwealth. In all willing it is absolutely a question of commanding and obeying, on the basis as already said, of a social structure composed of many 'souls'. Hence a philosopher should claim the right to include willing as such within the sphere of morals – morals being understood as the doctrine of the relations of supremacy under which the phenomenon of 'life' comes to be.[45]

The stability of the self is a provisional dynamic accomplishment resulting from negotiations among sub-personal drives (*Triebe*), here represented as citizens. One is internally complex, but one identifies with the dominant forces, or what James would call one's *habitual* centre of energy. The self for Nietzsche is thus a social construction, not in the familiar sense of a construction created by society – as when one claims, for example, that gender roles are socially constructed – but in the sense of *an internally constructed society*. The self is a society of sub-selves that operates best when it expresses the governing class's highest principles. Indeed, Nietzsche here identifies moral philosophy itself as the 'doctrine of the relations' among such sub-selves.

Nietzsche also describes this process in aesthetic terms, as in his admonition in *The Gay Science* to 'give style to one's character':

> *One thing is needful.* – To 'give style' to one's character – a great and rare art! It is practiced by those who survey all the strengths and weaknesses of their nature and then fit them into an artistic plan until every one of them appears as art and reason and even weaknesses delight the eye. Here a large mass of second nature has been added; there a piece of original nature has been removed – both times through long practice and daily work at it. Here the ugly that could not be removed is concealed; there it has been reinterpreted and made sublime. Much that is vague and resisted shaping has been saved and exploited for distant views; it is meant to beckon toward the far and immeasurable. In the end, when the work is finished, it becomes evident how the constraint of a single taste governed and formed everything large and small.[46]

Notable here is the incorporative action that is a decisive characteristic of will-to-power. When a distinctive vision or drive governs one's self-styling, it reframes other drives in a way that suits it. This provides an existential analogue to the Darwinian conception of *exaptation*, where extant structures gain new evolutionary functions over time.[47] The meaning or function of a structure is never settled by past uses. Indeed, because the interplay of drives has no natural finishing point, it provides continual work for one's form-giving powers. One literally embodies an ongoing moral and artistic project.

James and Nietzsche begin to come apart when one examines their evolutionary and physiological logics. Following Darwin's conception of non-directed variation (as adapted through Peirce's tychism), James holds that mental processes are inflected by a kind of randomness that irrupts throughout the natural world. An ineliminable accidental element is thus present in thought. Such accidental variation is inchoate and meaningless until it is sent out to construct the internal (habitual) and external (social-natural) environments. This may occur volitionally or by way of unconscious dynamics. Through this mediation of mental variation a centre of energy may be dislodged or promoted to a higher station in the habitual structure constituting the self. In this way the individual utilizes variational raw material to change its own future conditions for action. James thus uses Darwinian concepts to construe ontogeny as a nonlinear process that is mediated purposively by an evolved will, which inheres in higher-level systems such as society that may be construed on analogy with such ontogeny.[48]

Nietzsche's putative Darwinism or anti-Darwinism is the subject of several books and numerous articles.[49] What is clear is that Nietzsche accepts descent with modification; that he invokes the inheritance of acquired characteristics, which is not incompatible with giving some role to natural selection; and that he read very little by Charles Darwin himself, whom he tended to lump in with Spencer as an extreme externalist. Thus, while Nietzsche joins James in criticizing Spencer's passive construal of the individual, he levels this exact critique at Darwin: 'The influence of "external circumstances" is overestimated by Darwin to a ridiculous extent: the essential thing in the life process is precisely the tremendous shaping, form-creating force working from within which *utilizes* and *exploits* external "circumstances"'.[50] Nietzsche therefore does not follow James in using Darwinian non-directed variation as ammunition against externalism. On the contrary, Nietzsche's reason for critiquing Darwinism is the same as James's reason for celebrating it: a belief that agency must be reconstructed, not erased, in the science of life.

Nevertheless, Nietzsche did posit a quasi-Darwinian 'struggle for existence' *within the individual*.[51] Here he drew upon embryologist Wilhelm Roux's *Der Kampf der Theile im Organismus* ('Struggle of the Parts in the Organism') (1881). A disciple of influential German naturalist Ernst Haeckel, Roux had argued that tissues compete by vying for finite resources during ontogeny. Roux's theory takes on an existential cast in Nietzsche's writings, where it is psychological drives that vie for a place in the developing organism. Roux therefore stands in the background of Nietzsche's conception of the internalized oligarchy. In this sense, Nietzsche joins James as an early importer of Darwinian concepts into psychology, even if he downplays natural selection in the phylogenetic sense proposed by Darwin himself.

Nietzsche's psychologizing of Darwinian concepts is more reductionist than James's. Nietzsche deconstructs the individual into a swarm of drives. In doing so, he is sceptical of top-down volition. Nietzsche's internal struggle for existence therefore seems to reduce the individual to drives with no central function that mediates them: The 'me' is an effect but not a cause; no one is steering the ship. Nietzsche's critiques of the unified self reflect a healthy desire to dispel illusions about the simplicity or transparency of willing, but they also make his apparently earnest advice about self-cultivation paradoxical. How does one shepherd one's drives if one is nothing more than the shifting sum total of the very drives being shepherded? In contrast, James does not talk about 'wills' in the plural, reserving the term for a distinct evolved function that biases attention and thus action. The Jamesian will is in this way a fulcrum of moral agency, distinct from the centres of energy that it mediates.

To a Nietzschean, James might seem to retain a vestige of a soul, or an Archimedean point of volition inserted arbitrarily into a sea of competing forces. To a Jamesian, however, Nietzsche might seem to be plumbing for ontological depth where there is none. James looks to the ontological middle. The lowest levels of psychological reality reveal only intrinsically meaningless variation. Nothing properly intentional or purposive appears at a level below that of the organism as such – even if different centres of energy constitute this purposiveness in different ways. In contrast, Nietzsche posits homuncular 'under-souls' that seem to have a kind of agency and reality that the individual as such lacks.

James and Nietzsche agree that dogmatists delude themselves by appealing to origins – whether in a priori reason, clear-and-distinctness or scripture – to secure a foundationalist pedigree for their views. Their respective critiques of origins differ, however. In a sense, James looks upward and forward while Nietzsche looks downward and backward. Nietzsche traces values downward into physiology and backward into history in order to show that everything modern Christian Europe values as good, true and beautiful comes from places that are arguably bad, false and ugly. Christianity itself is said to be a religion, not of love, but of resentment in which the weak have tricked the strong to treat them as equals. If Nietzsche thus *unmasks* origins by revealing sedimented layers of purposes,[52] then James *de-emphasizes* origins by construing them as inchoate fodder for present and future purposes.[53]

This difference of emphasis aside, both James and Nietzsche construe self-transformation as a matter of promoting the aspects of oneself with which one would like to identify. This process always involves sacrificing something for something else. As James remarks in the *Principles*, 'I am often confronted by the necessity of standing by one of my empirical selves and relinquishing the rest.... This is as strong an example as there is of that selective industry of the mind on which I insisted'.[54] James thus folds his account of self-transformation into his general selectionist account of the sensorimotor system and mind: Just as perception requires sensory selection and attention requires cognitive selection, ethical life proceeds by way of existential selection in which one takes sides with certain potentialities over others. One butchers myriad ideals in the process.

James and Nietzsche agree that such a sacrifice may be made gladly if it is made in the name of an ideal. While ideals are not pre-given in a post-Darwinian universe, they may be posited; and their literal *incorporation* is what makes life significant. Indeed, James answers the titular question of his essay 'What Makes a Life Significant' with the claim that life requires both *strength* ('virtue') and an *ideal*: 'The significance of a human life ... is thus the offspring of a marriage of two different parents, either of whom alone is barren. The ideals taken by themselves give no reality, and the virtues by themselves no novelty'.[55] An ideal on this view is an unrealized possibility that differs from the present, while 'virtues' include the strength (or energy or will-to-power) to put an ideal into practice. Both are necessary. To paraphrase Kant, an ideal without virtue is empty, whereas virtue without an ideal is blind. James even identifies *intelligence* as the ability to recognize novel ideals, as opposed to an unintelligent humdrum attitude that always clings to some familiar good.

Character ideals: masters and saints

If James and Nietzsche agree on the necessity of positing ideals, they differ on what the ideal character would look like. Nietzsche's second mention in *The Varieties of Religious Experience* comes at an interesting turning point, not just in the book, but in James's entire body of work. This is because this is where James finally shifts to a (tentatively) normative mode in his ethics. *The Principles of Psychology* and *Talks to Teachers on Psychology* contain a moral psychology but no prescriptive ethics. The 1891 essay 'The Moral Philosopher and the Moral Life' recasts ethical habit-breaking at a historical and quasi-theological level (as described in Chapter 5), but it posits no ideal other than accommodating as many of the demands of sentient beings as possible. The actual determination of the best ordering of demands – whether in an individual or in society – is left to the future. In the fourteenth lecture of the *Varieties*, however, James seems confident enough to claim that one character type on the whole is most satisfactory. The ideal character is filling in its outlines.

The saint

James describes his ideal character type in religious terms, calling this figure the 'saint'.[56] To be a saint in James's sense means to think and act from a habitual centre of energy that relates to the world 'spiritually', on a certain liberal construal of the latter term. Specifically, James defines the saint in terms of the following psychological propensities (paraphrased here):[57]

1   A sense that one's life is continuous with an ideal power.
2   A willing surrender of oneself to the ideal power's control.
3   A sense of freedom and liberation from the normal confines of the selfhood.
4   A shift toward sympathy and love for the other as opposed to narrow egoism.

These qualities capture a sense of immersion in a wider universe that lies beyond the fringes of articulable experience, while also reflecting James's newfound emphasis in the *Varieties* on a passive experience of trust or self-surrender. James admits that such characteristics can be construed secularly, as a feeling of the relationship between self-aware consciousness and a tendency toward self-realization inherent in the unconscious. Nevertheless, he prefers to leave open the possibility of a literal backdoor to a spiritual realm that opens up through the unconscious. The most important thing is that in practice the above psychological propensities are supposed to promote several kinds of practical consequence:[58]

a   Asceticism: Positive pleasure in sacrifice, as a measure of loyalty to a higher power.
b   Strength of soul: Narrow egoistic anxieties are replaced by new expanses of fortitude.
c   Purity: Increased sensitivity to spiritual discord.
d   Charity: Tenderness for fellow-creatures.

James recognizes that each of these tendencies can be taken to pathological extremes. However, he claims that in harmonious combination they define the ideal human being.

James here introduces Nietzsche's 'master' type from the *Genealogy of Morals* as a foil. In James's words, 'The most inimical critic of the saintly impulses I know is Nietzsche. He contrasts them with the worldly passions as we find these embodied in the predaceous military character, altogether to the advantage of the latter'.[59] This does not sound like a Jamesian saint. In fact, James believes that Nietzsche would find his saint to be not only disgusting but threatening: 'For Nietzsche the saint represents little but sneakingness and slavishness. He is the sophisticated invalid, the degenerate *par excellence*, the man of insufficient vitality. His prevalence would put the human type in danger'.[60] James justifies this interpretation with a quotation that he translates from Nietzsche's *Genealogy of Morals*:[61]

> The weaker, not the stronger, are the strong's undoing. It is not *fear* of our fellow-man, which we should wish to see diminished; for fear rouses those who are strong to become terrible in turn to themselves, and preserves the hard-earned and successful type of humanity. What is to be dreaded by us more than any other doom is not fear, but rather the great disgust, not fear, but rather the great pity – disgust and pity for our human fellows.[62]

This is a characteristically Nietzschean passage: The weak are said to be spiritually poisoned by their seething resentment, leading them to conspire against the strong in order to drain them of their strength and secure a relatively higher position of power. Although Nietzsche nowhere mentions saints in this passage, James believes that he would look upon the saint of the *Varieties* as an example of such despicable weakness.

The master

As in his conflation of Nietzsche with Schopenhauer, James fails to take note of his and Nietzsche's substantial areas of agreement. Most importantly, an expression of strength for Nietzsche is not to be equated simplistically with predation or physical violence. Although one of Nietzsche's most striking moves is to claim that the strong need to be defended from the weak, the primary meanings of terms like 'strong' and 'weak' in Nietzsche's writings are psychological or spiritual rather than physical. Will-to-power's expressions include spontaneous creative expression in play or art, a spirit of magnanimity and forgiveness stemming from a sense of one's own abundance and the expression of one's highest values through philosophy. Violence is the province, not of the master, but of the 'blond beast', which Nietzsche defines as a brutal and primordial type that lives on in some of us as a 'hidden core' of potential violence.[63] Although Nietzsche romanticizes the blond beast as a type that existed before the taming influence of culture, his point in discussing this figure is not that we should (*per impossible*) revert to a pre-cultural bestial state. Rather, it is that the image of a beast can remind us that it is better to be awed by something terrible than to live insipid or mediocre lives. A central task in interpreting Nietzsche then consists in determining the precise sense in which he wants us to be terrible, strong or noble, considering that he is not being straightforward in how he uses such terms.

Nietzsche's master or strong type is defined by spontaneous self-expression in 'vigorous, free, joyful' activity.[64] The master is automatically self-affirming. In contrast, the primary valuation of the weak is not the spontaneous 'I am good' but rather the reactive 'the strong are evil'. Nietzsche thus construes slave morality as a reaction against the strong by the weak. Based on such statements alone, Nietzsche seems to favour the unreflective expression of internal drives. However, Nietzsche also prizes wisdom, artifice and deliberate character-formation, as in his exhortation to give style to one's character. Nietzsche's ideal is not the unreflective brute but rather something more deliberate and artistic. In particular, Nietzschean strength is not paradigmatically other-directed, as in an expression of domination over another individual. It is cultivated, nuanced and expressed in the interplay of active and passive dimensions of individual experience. To be masterly for Nietzsche must be understood in terms of the interplay of spontaneous affirmation and deliberative self-cultivation within the individual. The question is when to modulate between these strategies – that is, when to intervene in order to encourage or suppress a drive that represents a particular way of being.

It is no different with James. The central ethical problematic for James is when to favour an idea by attending to it wilfully and when to allow automatic unconscious forces to hold sway. In a sense the former mode is active and the latter passive, but the story is more complicated than this. The will is active vis-à-vis cognitive variation, which it selects upon to produce associated motor consequences; and yet it is passive vis-à-vis this same variation in that it cannot directly produce or elicit it. The will also 'passively' lies dormant during habitual ideo-motor action, although this may be a studied dormancy that demonstrates faith in one's own unconscious (and perhaps spiritual) dynamics. One lays down one's arms in ceasing the repression of unknown powers. Different aspects of oneself are active or passive relative to one another, in multiple senses of activity and passivity.

For both James and Nietzsche, therefore, the generation of internal resistance becomes an art and a practice. Both therefore recommend a kind of measured asceticism. This is not to prescribe asceticism itself as an ideal, as if one could live on negations alone. It is only to recommend asceticism as one dynamogenic agent among others. Asceticism is necessary in self-transformation, as here one aims precisely to deny oneself the impulses that come the easiest. It requires discipline to deny momentary satisfactions in favour of long-term ones. One may even learn to enjoy the flouting of immediate enjoyments if one identifies powerfully with the part of oneself doing the flouting. Challenges as such may then become indirect sources of gratification, as in Nietzsche's famous (if hyperbolic) maxim, 'what does not destroy me makes me stronger'.[65]

James is no less severe than Nietzsche on this front, including asceticism as one of the traits typical of the saint. Asceticism is also closely allied to the saintly characteristic of purity, which makes the saint 'exceedingly sensitive to inner inconsistency or discord'.[66] Self-transformation is after all a form of sacrifice. James even goes as far as to define 'weakness of character' as the inability to target one's 'inferior self and its pet softness'.[67] This kind of self-severity is far afield from the stereotype of pragmatism as facile wishful thinking.

Nevertheless, Nietzsche is a greater theorist of asceticism than James, devoting the entire third essay of the *Genealogy* to a survey of asceticism's expressions. Nietzsche's position is not that asceticism is absolutely good or bad but that it provides a lens through which to critique a wide variety of cultural practices. Asceticism is fascinating to Nietzsche because it raises deep questions about his hypothesis of will-to-power. If life is will-to-power, then asceticism is paradoxical. How can a will toward expansion and incorporation revel in its own denial and contraction? The answer must be that will-to-power serves itself indirectly through finding meaning in its negations, savouring its own capacity for turning back upon itself. The ascetic dimension of will-to-power is thus tied to its function of constructing and commanding, as it seeks to impose order and not merely to play itself out blindly. Nietzsche is clear that the imposition of constraint is a positive function of will-to-power that generates meaning. Nietzsche is concerned, however, by the nihilistic logic of a will-to-power that has no end other than to will nothingness.

Interestingly, Nietzsche describes the very drive for truth as a form of asceticism. The drive for truth is ascetic for Nietzsche insofar as it signals a desire for confrontation with a recalcitrant reality that exists independently of one's own capacities or wishes. A central theme of Nietzsche's thought is that knowledge may be dangerous and even antithetical to life. This means that a wholly honest philosophy may require an entirely new kind of being – such as Nietzsche's *Übermensch* – in order (quite literally) to be incorporated. It may be unpleasant to suppose, for example, that there is no hidden benevolent presence in the universe or that our highest values derive from unsavoury unconscious impulses. The ascetic dimension of epistemology is taken to an extreme by Vedic and Kantian philosophies that posit a reality beyond the veil of sense about which nothing articulate can be said. In Nietzsche's words,

> To renounce belief in one's ego, to deny one's own 'reality' – what a triumph! not merely over the senses, over appearance, but a much higher kind of triumph, a violation and cruelty against *reason* – a voluptuous pleasure that reaches its height when the ascetic self-contempt and self-mockery of reason declares: '*there is* a realm of truth and being, but reason is *excluded* from it!'[68]

When philosophy, or the 'love of knowledge', becomes an exercise in limning the severe limits of knowledge, then something ascetic is going on. The true nature of both the knower and the known may remain a mystery. Nietzsche's point is not to subscribe to such a view but to point out how philosophy can take the form of a refined self-cruelty. From a certain perspective, Kant's first *Critique* might as well have been titled *The Self-Mockery of Reason*.

A cooperative universe

James gets Nietzsche wrong in fundamental ways in the *Varieties*: Nietzsche is *not* a cantankerous pessimist who is unconcerned with redeeming existence from suffering; and he does *not* identify strength with the physical domination of the weak. Like James, his entire philosophy is a sustained attempt to affirm agency, where the crucible of agency is a multivalent and dynamic self. Nietzsche is thus much more Jamesian than James seems to realize. Nevertheless, James and Nietzsche have real differences. In the final analysis, James's saint and Nietzsche's master belong to different moral and social visions. This can be demonstrated starting with an apparent about-face in James's attitude toward Spencer.

As described above, James mocks Spencer's utopian vision in much the same manner as Nietzsche. It would be simplistic, however, to claim that James was wholly against Spencer.

James was strongly influenced by Spencer in his youth, when he subscribed to receive his works in serial instalments.[69] James was particularly taken by Spencer's view that all aspects of reality – cosmic, biological, psychological and societal – are evolving. In James's words, 'Spencer was the first to see in evolution an absolutely universal principle.'[70] Spencer also introduced the modern ecological conception of 'environment' to the English-speaking world, generating a conceptual dyad in which the organism is tethered to a unified set of circumstances.[71] This provides a powerful framework: Everything is evolving, and evolution is always relative to an environment. This environment-relativity includes psychological and social evolution, such that even knowledge and values are relative to environments. The conclusion is that there is no absolute knowledge or values. Spencer thus posits in a quasi-Kantian fashion that beyond relative knowledge there is only the 'Unknowable'.

James does not follow Spencer in positing an unknowable realm. He also challenges Spencer's view that the organism-environment relation is a one-sided affair in which the former passively adapts to the latter, as well as Spencer's understanding of the world's inexorable progress toward a perfect state of biological and social adaptation. The part of Spencer that James retains as central to his philosophy is the view that individuals and their values must be understood in terms of their function in society. There is no such thing as an ideal character independent of any environment. Such a view may not sound sufficiently 'individualistic' for James, given the amount of stress that has been placed on the individual throughout the present study. However, just because James emphasizes the efficacy of the individual in mediating itself and its embedding systems does not mean that the individual's own value does not in some way derive from its functioning within these systems. Nothing has a value or a function in a vacuum.

James identifies this conception of the environment-relativity of values with Darwin as well as Spencer.[72] Nevertheless, Spencer's published evolutionism predates Darwin's, and Spencer is James's reference for the relativity of values in the *Varieties*. Strikingly, in fact, James cites the methodology of Spencer's *Data of Ethics* – the very work that he had parodied earlier – in outlining his own utopian vision of sainthood:[73]

> I think that the method which Mr. Spencer uses in his *Data of Ethics* will help to fix our opinion. Ideality in conduct is altogether a matter of adaptation.... It is meanwhile quite possible to conceive an imaginary society in which there should be no aggressiveness, but only sympathy and fairness – any small community of true friends now realizes such a society. Abstractly considered, such a society on a large scale would be the millennium, for every good thing might be realized there with no expense of friction. To such a millennial society the saint would be entirely adapted.[74]

Two points must be taken away from this passage. The first is that a character ideal for James implies an ideal of a character-in-a-society. Because '[i]deality in conduct is altogether a matter of adaptation', any ideal character implies a social environment in relation to which it is ideal.[75] James's ideal is therefore not the saint per se but the community-of-mutually-adapted-saints. Indeed, for reasons that have now been formalized by game theory, there is risk in exhibiting the openness and vulnerability of the saint in a world consisting of a diverse mixture of character types. James's ideal of the community of saints is thus only an ideal 'abstractly considered'. It therefore serves as a *regulative ideal*, much like what James calls 'the most inclusive realizable moral whole' in 'The Moral Philosopher and the Moral Life' or 'absolute truth' in *Pragmatism* and *The Meaning of Truth*. That is, it provides a hypothetical vanishing point toward which to strive. James does not expect the ideal society to be realized soon, or perhaps ever. To the extent that we can muster the requisite 'strenuous mood', however, we plant the seeds of a better and more saintly future.[76] James thus claims that the function of the saint-like among us today is to orient us toward the ideal, as 'impregnators of the world, vivifiers and animaters of potentialities of goodness'.[77] This may be disastrous to a saint's *biological* success – as in the case of martyrdom – while providing useful ideal novelty in social or historical evolution.

The second point is that James has reversed the opinion, expressed in the voice of the 'subjectivist' in 'The Dilemma of Determinism', that suicide would be preferable to a world without conflict. As against this early critique, James now concedes that a world of peace is a worthy goal. James's considered opinion is that such a world may provide grist for the moral mill, after all. As James argues in 'The Importance of Individuals' (1890), even small differences between individuals may be cause for controversy and excite great strenuous vigour. Indeed, the smaller the differences, the greater the tendency to magnify them in order to make them significant. This is not a deceptive exercise, as it is the interests of sentient beings that make anything interesting in the first place. Nothing is absolutely significant or insignificant. This means that there is no danger that a Spencerian – or, as it now may be said, *Jamesian* – utopia of gentleness will evaporate all significance from the world. There is always an *active zone* of interest around current struggles, although it may narrow over time:

> And though it may be true, as Mr. Spencer predicts, that each later zone shall fatally be narrower than its forerunners; and that when the ultimate ladylike tea-table elysium of the *Data of Ethics* shall prevail, such questions as the breaking of eggs at the large or the small end will span the whole scope of possible human warfare ... what eagerness there will be![78]

James addresses this topic with characteristic humour, but his point is serious: The elimination of irrational violence or prejudice from the world is not a barrier to moral significance. The resistance that gives us meaning is not going away. Resistance is inherent to life, whether it is provided by personal projects, good-natured rivalry or simply by the inherent tension of being a habitually constituted organism attempting to chart a course from the myopic egotism of childhood to something more expansive and humane.

This is the proper context for understanding James's concept of 'a moral equivalent of war'. Although this concept appears in James's 1910 essay of the same name, he actually introduces it in his discussion of saintliness in the *Varieties* to mark the need for 'something heroic that will speak to men as universally as war does, and yet will be as compatible with their spiritual selves as war has provided itself to be incompatible'.[79] James's suggestion – which helped to inspire both the Depression-era Civilian Conservation Corps and the Peace Corps in the US – is that military conscription should be replaced by 'conscription of the whole youthful population to form for a certain number of years a part of the army against *nature*'.[80] James argued that the enlistment of youth into civic projects would have multiple benefits: The public good would be served directly through the creation of infrastructure and other goods; and the privileged would form bonds of empathy and friendship with the working classes and the poor, thus countering the systematic blindness toward the interests of others that James had articulated poetically in his prior essay 'On a Certain Blindness in Human Beings'.[81]

In other words, James believes in a genuinely cooperative universe in which the value of the individual is determined, at least in part, by its role in creating a just society. Nietzsche may not have advocated for brutality, but a final comparative analysis does show him to be less concerned than James with the convergence of ideals on a widely shared state of flourishing.

Two thought experiments

Another of James and Nietzsche's commonalities is that they were among the finest literary talents of the philosophical tradition. Both were thus adept at illustrating their ideas through creative metaphors, analogies and thought experiments. A comparison of two such thought experiments will provide a useful coda to the present discussion.

The first is Nietzsche's thought experiment in which he presents his doctrine of the *eternal recurrence*, or the infinite repetition of the entire history of the universe. In a well-known passage in *The Gay Science*, Nietzsche enlists a fictional demon to test the reader's reaction to the idea of repeating the same life in all its details:

> Would you not throw yourself down and gnash your teeth and curse the demon who spoke thus? Or have you once experienced a tremendous moment when you would have answered him: 'You are a god and never have I heard anything more divine.' If this thought gained possession of you, it would change you as you are or perhaps crush you. The question in each and every thing, 'Do you desire this once more and innumerable times more?' would lie upon your actions as the greatest weight. Or how well disposed would you have to become to yourself and to life *to crave nothing more fervently* than this ultimate eternal confirmation and seal?[82]

Nietzsche seems to have held the eternal recurrence as a genuine metaphysical thesis, which hit him with the force of a revelation in the Swiss Alps in 1881. Regardless, the idea also works on a hypothetical level as a measure of one's ability to affirm existence: To what extent do you affirm life in all of its aspects, including its torments as well as its pleasures? The question is a test of one's strength in the actual Nietzschean sense of the term, which is the ability to affirm even the ugliest parts of existence.

James makes a similar rhetorical gesture in *Pragmatism*, proposing an encounter with a supernatural entity that is soliciting a reaction about one's willingness to participate in a particular kind of world. For James, the question is whether to take part in an *uncertain* universe:

> Suppose that the world's author put the case to you before creation, saying: 'I am going to make a world not certain to be saved, a world the perfection of which shall be conditional merely, the condition being that each several agent does its own "level best". I offer you the chance of taking part in such a world. Its safety, you see, is unwarranted. It is a real adventure, a real danger, yet it may win through. It is a social scheme of co-operative work genuinely to be done. Will you join the procession? Will you trust yourself and trust the other agents enough to face the risk?'
>
> Should you in all seriousness, if participation in such a world were proposed to you, feel bound to reject it as not safe enough? Would you say that, rather than be part and parcel of so fundamentally pluralistic and irrational a universe, you preferred to relapse into the slumber of nonentity from which you had been momentarily aroused by the tempter's voice?[83]

For James as for Nietzsche, to answer the mythical being in the affirmative is to exhibit the hardihood necessary to live a meaningful existence. Neither one's individual life nor life in general is logically determined to culminate in anything meaningful. Salvation depends on what we do.

Rhetorical similarity aside, these two thought experiments are asking two very different things. First, they represent opposed views about the nature of time and progress. To be sure, both James and Nietzsche reject the view – shared in different forms by Spencer and the absolute idealists – that history perfects itself on logical or metaphysical principle. However, James retains a linear view of history, whereas Nietzsche does not. According to James's meliorism, individuals progress in piecemeal fashion toward an inchoate ideal of moral inclusiveness and absolute truth. This is not Victorian progress as if on rails (unless here the tracks are built while the train is in motion), but James does draw a line from the present toward a future vanishing point. In contrast, Nietzsche draws a circle, reflecting an Eastern-tinged fatalism on which existence is always-already completed. Indeed, according to Nietzsche's theory, one's choices are not only going to be repeated infinitely in future cycles; they have already been repeated an infinite number of times in past ones. In short, James is asking his reader to affirm *possibility*, while Nietzsche is asking his reader to affirm *necessity*.

James's deepest metaphysical commitment is to pluralism. This is not just the epistemological view – shared by Nietzsche – that no standpoint captures the full reality of any given object. It is also the metaphysical thesis that determinism and fatalism are literally false and that nature does not exist in any totalized or 'rounded-out' fashion. A pluralistic universe is loose in its joints, through which novelty flows continuously. Even if the universe were to be reborn on James's view, it would be different with each cycle. Nietzsche's doctrine of the eternal recurrence denies this. Nietzsche depicts a world in which competing forces are continually exerting themselves to their limit, with a fatalistic (if complex and highly interpretable) result that is repeated infinitely. Nietzsche's fatalism is captured by his stoical concept of *amor fati* ('the love of fate'): The highest yes-saying is the affirmation of what is true in any event. This again presents a paradoxical aspect of Nietzsche's philosophy, which in some places emphasizes initiative and choice. Perhaps Nietzsche is attempting to supersede the distinction between freedom and necessity in a compatibilist fashion.[84] Perhaps freedom is just a certain way of inhering in a deterministic matrix, or the very concept of determinism as a restriction on freedom is a verbal trick. For James, however, the distinction between a monistic universe and a pluralistic one is genuine, not an illusion or the result of a falsely reified dichotomy.

Second, as already suggested above, James values empathy and social cohesion more than Nietzsche does. James describes his pluralistic universe explicitly as 'a social scheme of co-operative work genuinely to be done'. He also argues that the value of an individual is at least partially a function of its society, where the ideal society excludes pernicious class distinctions. Nietzsche does not concede society such a power to determine the value of its inhabitants, nor does he believe that hierarchies of value should be maximally inclusive of the interests of those they affect. To Nietzsche this would be a 'levelling of the mountains' that enforces a mediocre existence in the name of equality. This is not to subscribe to the caricature of Nietzsche as a monster, but only to acknowledge the depth of his anti-democratic and even misanthropic tendencies. James was a popularizer who believed in the populace; Nietzsche was anything but. This difference is reflected in the different expectations they held about reactions to their respective thought experiments. James claims that the majority will be up to the collective task of constructing a better world. In contrast, Nietzsche thinks that a scarce few could stomach his more individualistic affirmation of existence, but that this very scarcity makes them precious. To speak more biographically – as both James and Nietzsche might have done – James was a famously warm and gregarious person who believed in pluralistic democracy, whereas Nietzsche was an isolated elitist who disparaged everything common.

This context sheds light on James's and Nietzsche's respective historical roles as defender of religion and God's coroner. As described in Chapter 5, James's depiction of linear melioristic progress takes on a theological cast. In particular, James argues that there may be a hierarchy of increasingly inclusive consciousnesses, where the widest mind represents the divine culmination of all sentient life. This is the dizzying cybernetic outer limit of James's psychological selectionism: God as the highest level in a hierarchy of selves that each filter material from the levels beneath. This image of the individual as part of a finite and growing divine being is meant to add intensity and urgency to individual moral action.

If James posits such a finite God in order to encourage 'the strenuous life', then Nietzsche refuses humanity such a divine presence for similar reasons. One gains a certain energy from looking hopefully upward, but one also generates a certain intensity from inward-looking self-overcoming. This is a difference of degree, as both James and Nietzsche prize ascetic self-overcoming. The point is that Nietzsche is more the connoisseur of asceticism and its fruits. Nietzsche thus posits the *Übermensch*, which is like the master of the *Genealogy of Morals* but to a higher degree. Both the *Übermensch* and the eternal recurrence are described rhapsodically in Nietzsche's philosophical novel *Thus Spoke Zarathustra*. The *Übermensch* says 'yes' to the eternal recurrence, while eschewing God or any of the other metaphysical trappings undergirding the philosophical tradition from Plato to Christianity. Notably, the *Übermensch* is essentially 'this-worldly' in its refusal to denigrate empirical reality by comparison to something purportedly ideal or divine. James's heterodox spirituality may be this-worldly as religion goes, with its focus on concrete practice, but it is premised on a certain indulgence of improvable metaphysical hopes. Nietzsche does not encourage willing-to-believe one's hopes so much as willing-to-know harsh truths.

Thus, for all of their similarities, James and Nietzsche have different perspectives on optimal human (or post-human) functioning: James wagers that individuals will always flourish best when strenuously contributing to an imagined spiritual 'more' that engulfs us; Nietzsche's experiment is to invoke a post-human figure that thrives best precisely in conditions that frustrate such metaphysical needs. If Nietzsche is correct that truth-seeking is ascetic, then he envisions a hardened asceticism that savours the truth of a godless cyclical world in which myriad wills expand and frustrate themselves toward no convergent end.

Conclusion

Working independently, James and Nietzsche each proffered an ethics of self-transformation developed in explicit opposition to Schopenhauer's 'cantankerous' pessimism and Spencer's asymmetrically externalist psychology and biology. Both thinkers fundamentally agree that a meaningful life must employ the energetic and form-giving powers of the individual. Challenge is in this sense a positive thing. In James's words, 'the world is all the richer for having the devil in it, *so long as we keep our foot upon his neck*'.[85] Moreover, both construe the individual as a multi-valent organism composed of hierarchically ordered sub-selves, and both consider the mediation of this structure to be the proper subject of moral philosophy.

James and Nietzsche are reacting to the Darwinian subversion of traditional images of human nature, and both do so by rejecting traditional metaphysics as well as reductionistic materialistic science. However, whereas James utilizes Darwinian conceptual tools to reconstruct individual agency and purpose, Nietzsche has a more complex and antagonistic relationship with Darwinism. If self-transformation for James means wilfully mediating non-directed variation to alter one's habitual structure, self-transformation for Nietzsche means a 'struggle for existence' among under-souls that may not be mediated by any centralized power. These under-souls *are* will-to-power and therefore express the tendency toward expansion and incorporation that characterizes all of life and nature.

James does not set a good example by reading Nietzsche simplistically as a proponent of violence. Action-oriented philosophies like James's and Nietzsche's must be situated within a framework of intelligence and moral reflection. The danger of an irrational philosophy of action can be seen in the actual co-optation of both James and Nietzsche by right-wing totalitarian states during the twentieth century. The National Socialists' co-optation of Nietzsche – who criticized both anti-Semitism and German nationalism – is well known. Less well known is Benito Mussolini's fandom of both James and Nietzsche. Mussolini praises both James and Nietzsche, for instance, in a 1926 interview:

> Nietzsche enchanted me when I was twenty, and reinforced the anti-democratic elements in my nature. The pragmatism of William James was of great use to me in my political career. James taught me that an action should be judged by its results rather than by its doctrinary basis. I learnt of James that faith in action, that ardent will to live and fight, to which Fascism owes a great part of its success.[86]

The point here is not that philosophers should be held responsible for the most irresponsible uses of their views. It is that a philosophy of action must not lose sight of rationality. Thought exists for the sake of action, as in the doctrine of the reflex arc. Given that the arc is really a circuit, however, action also exists for the sake of thought and is properly guided by it.

In the final analysis, James seeks to enact a utopia that is hospitable to humanity's extant social and metaphysical needs, while Nietzsche invokes an exemplary post-humanity that digests those same needs through ascetic self-overcoming. If the danger of the former view is wishful thinking, the danger of the latter view is alienation from ourselves and others. The choice is a pragmatic one with no pre-given answer.

Notes

1 For James on temperament, see Bordogna (2001).
2 For James and disciplinary boundaries, see Bordogna (2008).
3 Nietzsche 1968a/1887 I, §17.
4 James similarly outlines an interdisciplinary 'science of man' in proposing to teach the first undergraduate physiological psychology in the US, as described in Chapter 3.
5 Richards 1987, 415.
6 VRE, 28–29.
7 Nietzsche 1974/1882, Preface, §3.
8 The evidence is adduced by Thomas Brobjer: Brobjer 2004, 45; 2008, 103–104.
9 Joly 1883, chap. 3. Joly cites James's (1881) French article 'Les grands hommes, les grandes pensées et le milieu'.
10 A longer study of James and Nietzsche would need to examine the influence of Emerson. Although Emerson was a James family friend and James's godfather, he may have been a stronger influence on Nietzsche than James. Emerson is practically the only thinker to whom Nietzsche gives unqualified praise. See Stack (1992) and Ratner-Rosenhagen (2011).
11 James's extant library at Harvard contains second editions of *Jenseits von Gut und Böse* and *Zur Genealogie der Moral*.
12 Thanks to Megan Mustain for pointing out several of the pieces of correspondence cited here.
13 Émile Faguet's 1903 *En lisant Nietzsche* (CWJ 10, 567); Daniel Halévy's 1909 *La vie de Frédéric Nietzsche* (CWJ 12, 510).
14 Salter 1917. James writes to E. Gibbons that he anticipates Salter's finishing his work on Nietzsche (which was published several years after his death) (CWJ 12, 540). Two preceding works on Nietzsche in the US are Dolson (1901) and Mencken (1908).
15 CWJ 12, 540.
16 CWJ 12, 265–266.
17 CWJ 12, 533.
18 CWJ 12, 538; 527.
19 CWJ 12, 373.
20 In an essay on Italian pragmatism, James mentions Nietzsche in a list of philosophers discussed by Giovanni Papini (EPH, 145); Papini also seems to have reminded James of Nietzsche, since James wrote 'Nietzsche's tho't as well as style' on a copy of an article by Papini (EPH, 244). In another essay, James compares the attitude of mystic Benjamin Blood with Nietzsche's idea of *amor fati* ('love of fate'), thus showing a familiarity with one of Nietzsche's terms of art (EPH, 189). Finally, in his 'Notes on Ethics' from 1899–1901, James depicts Nietzsche as sickly and peevish (MEN, 313). This is in line with James's portrayal of Nietzsche in the *Varieties* discussed here.
21 See the essays of Pütz (1995).
22 ECR, 508.
23 Despite the proliferation of such studies as *Freud and Nietzsche* (Assoun 2000) and *Wittgenstein and William James* (Goodman 2002) – and even a book comparing Nietzsche with James's novelist brother Henry (Donadio 1978) – there is at time of press no book on James and Nietzsche in English. For a German study, see Hingst (1998); for a French one, see Karakas (2014). Shorter comparative discussions tend to focus on truth. This is the case in Allen (1993, chaps. 3–4; 1994) and Cormier (2001,

chaps. 1–2). Richard Rorty enlists both James and Nietzsche as allies but does not offer detailed comparisons. However, see Rorty (1982) and Rorty (2007), and see Boffetti (2004) for a critique of Rorty's Nietzschean reading of James. Closer to the present study are the James-Nietzsche comparisons of Franzese (2003; 2008, 194–200), which are centred on philosophical anthropology rather than truth.

24 VRE, 34.

25 VRE, 39.

26 Perry 1935 I, 723–724. This version of the letter appears not to have been sent.

27 See the 'Attempt at a Self-Criticism' that Nietzsche appended to the second edition of *The Birth of Tragedy*: 'How far removed I was from all this resignationism!' (1968b/1872: §6). For the yes-saying of the saint, see VRE (219–220).

28 Cited in Richards (1987, 426–427).

29 Nietzsche 1968a/1887 II, §15.

30 Nietzsche 1968a/1887 I, §10.

31 VRE, 240.

32 WB, 130.

33 WB, 130–131.

34 Nietzsche 1974/1872, §373. Walter Kaufmann draws attention to the parallel between James's and Nietzsche's attitudes toward Spencer in his translation of the latter passage (note 135).

35 Nietzsche 1974/1882, §373.

36 For more on Nietzsche and Spencer, see Call (1998) and Moore (2002a).

37 ERM, 129–146. See also 'The Powers of Men' (ERM, 147–161).

38 ERM, 145.

39 Rolph 1882; Moore 2002b, 47.

40 Nietzsche 1966/1886, §36.

41 Nietzsche's radically interpretive orientation prefigures the French poststructuralist movement that he influenced starting in the early 1960s. See Hoy (2004, 19–56).

42 MEN, 314.

43 Nietzsche 1968a/1887 I, §17.

44 Nietzsche 1968a/1887 II, §1.

45 Nietzsche 1966/1886, §19.

46 Nietzsche 1974/1882, §290.

47 Gould and Vrba 1982.

48 James even describes reality as a whole on analogy with ontogeny, as described in Chapter 5.

49 As for the books, Moore (2002) undermines the myth of Nietzsche's 'untimeliness' by situating his work in terms of his biological influences; Richardson (2004) provides a naturalistic reading of Nietzsche that construes will-to-power as a product of natural selection rather than an ontologically primordial force; and Johnson (2010) reads Nietzsche's *Genealogy of Morals* as a sophisticated critique of Darwinism. See also Düsing (2006).

50 Nietzsche 1968c, §657.

51 Müller-Lauter 1999, 162–164; Moore 2002.

52 Nietzsche practices what Williams (2004) calls 'unmasking genealogy', or the criticism of a cultural practice by way of an investigation of its origins. See also Hoy (2009, 224–242).

53 It is tempting to claim that Nietzsche represents a Continental tendency to invoke internal forces and structures (as in Kantianism, vitalism and rational morphology), whereas James represents an anglophone emphasis on external forces (as in empiricism, Spencer's psychology and biology, psychological behaviourism and neo-Darwinism). Indeed, James saw himself as embracing anglophone philosophical attitudes, while Nietzsche practically equated 'English' with 'superficial' and stood closer to vitalism and neo-Kantianism. However, all of this is complicated by

Nietzsche's radical interpretivism, as well as by James's anti-externalism and his increasingly Bergsonian view of the surging 'creativity' of nature (discussed in Chapter 5).

54 PP I, 296.
55 TT, 164.
56 VRE, 219.
57 VRE, 219–220.
58 VRE, 221. This list is again paraphrased.
59 VRE, 295.
60 VRE, 296.
61 VRE, 297. The quotation is from the third essay of the *Genealogy of Morals* (§14). James's translation is loose but does not change the meaning of the passage. James admits to having 'abridged, and in one place transposed, a sentence' (VRE: 297n24). In fact James omits two sizeable chunks of text, only one of which he indicates with an ellipsis. James's quotation is long and is not given in full here.
62 VRE, 296–297.
63 Nietzsche 1968a/1887 I, §11.
64 Nietzsche 1968a/1887 I, §7.
65 Nietzsche 1954a/1888, 'Maxims and Arrows', §8. Nietzsche puts this remark in the first person, leaving it open whether he means it to apply to others. In any event, it did not apply even to Nietzsche.
66 VRE, 234.
67 VRE, 214.
68 Nietzsche 1968a/1887 II, §12.
69 EPH, 116.
70 EPH, 114.
71 Pearce 2010.
72 James scholar Ignas Skrupskelis (2007) has unearthed a long-missed 1883 letter on this matter from James to Charles Darwin's son William Erasmus Darwin:

> Now I take it that your father meant to protest against this ideal of a perfection equally binding on all types of creature.... Into the question of the comparative rightness of the 'ways' or that of the comparative perfection of the several types as measured by an absolute standard, a question which any professional moralist would then immediately take up, your father does not enter. Of course he would have said here too that the notion of an absolutely perfect type, considered out of connexion with 'conditions of existence' was a meaningless idea to him.
>
> (750)

The James family was acquainted with the Darwins from the late 1860s (Taylor 1990).
73 Indeed, James claims in a tribute to Spencer that *The Data of Ethics* is his greatest work and that it is bound to be an enduring classic because it is a vital expression of one of the great philosophical attitudes toward life (EPH, 100).
74 VRE, 298.
75 Franzese (2008) claims that James's criticism of Nietzsche in the *Varieties* is just this: James is a relativist about character types but Nietzsche is not (194–200). This is indeed an important part of James's criticism.
76 VRE, 298–299.
77 VRE, 285.
78 WB, 193.
79 VRE, 292.
80 ERM, 171.
81 TT, 132–149.
82 Nietzsche 1974/1882, §341.

83 P, 139.
84 James employed the term 'soft determinism' in rejecting such a view. See 'The Dilemma of Determinism' (WB, 114–140).
85 VRE, 48.
86 Quoted in Perry (1935 II, 575). Notably, Mussolini claims that his favourite philosopher is neither James nor Nietzsche but Georges Sorel.

References

Allen, B. 1993. *Truth in Philosophy*. Cambridge, MA: Harvard University Press.
Allen, B. 1994. Work on Truth in America: The Example of William James. In *Cohesion and Dissent*, eds C. Colatrella and J. Alkana, 95–113. Albany, NY: State University of New York Press.
Assoun, P. 2000. *Freud and Nietzsche*. Translated by R. L. Collier. London: Athlone Press.
Boffetti, J. M. 2004. Rorty's Nietzschean Pragmatism: A Jamesian Response. *Review of Politics* 66, no. 4: 605–631.
Bordogna, F. 2001. The Psychology and Physiology of Temperament: Pragmatism in Context. *Journal of the History of the Behavioral Sciences* 37, no. 1: 3–25.
Bordogna, F. 2008. *William James at the Boundaries: Philosophy, Science, and the Geography of Knowledge*. Chicago, IL: University Of Chicago Press.
Brobjer, T. H. 2004. Nietzsche's Reading and Knowledge of Natural Science: An Overview. In *Nietzsche and Science*, eds G. Moore and T. H. Brobjer, 21–50. Aldershot: Ashgate.
Brobjer, T. H. 2008. *Nietzsche's Philosophical Context: An Intellectual Biography*. International Nietzsche Studies. Edited by R. Schacht. Urbana, IL: University of Illinois Press.
Call, L. 1998. Anti-Darwin, Anti-Spencer: Friedrich Nietzsche's Critique of Darwin and 'Darwinism'. *History of Science* 36, no. 1: 1–22.
Cormier, H. 2001. *The Truth is What Works: William James, Pragmatism, and the Seed of Death*. Lanham, MD: Rowman & Littlefield.
Dolson, G. N. 1901. *The Philosophy of Friedrich Nietzsche*. New York: The Macmillan Company.
Donadio, S. 1978. *Nietzsche, Henry James, and the Artistic Will*. New York: Oxford University Press.
Düsing, E. 2006. *Nietzsches Denkweg: Theologie, Darwinismus, Nihilismus*. Munich, Germany: W. Fink.
Faguet, E. 1903. *En lisant Nietzsche*. Paris: Société Française d'Imprimerie et de Librairie.
Franzese, S. 2003. James Versus Nietzsche: Energy and Asceticism in James. *Streams of William James* 5, no. 2: 10–12.
Franzese, S. 2008. *The Ethics of Energy: William James's Moral Philosophy in Focus*. Frankfurt, Germany: Ontos Verlag.
Goodman, R. B. 2002. *Wittgenstein and William James*. Cambridge: Cambridge University Press.
Gould, S. J. and E. S. Vrba. 1982. Exaptation-A Missing Term in the Science of Form. *Paleobiology* 8, no. 1: 4–15.
Halévy, D. 1909. *La vie de Frédéric Nietzsche*. Paris: Calmann-Lévy.
Hingst, K. M. 1998. *Perspektivismus und Pragmatismus: Ein Vergleich auf der Grundlage der Wahrheitsbegriffe und Religionsphilosophien von Nietzsche und James*. Würzburg, Germany: Königshausen und Neumann.

Hoy, D. C. 2004. *Critical Resistance: From Poststructuralism to Post-Critique.* Cambridge, MA: MIT Press.

Hoy, D. C. 2009. *The Time of Our Lives: A Critical History of Temporality.* Cambridge, MA: MIT Press.

James, W. 1881. Les grands hommes, les grandes pensées et le milieu. *Critique Philosophique* January–February 1881.

James, W. 1975. *Pragmatism.* The Works of William James. Edited by F. Burkhardt, F. Bowers and I. K. Skrupskelis. Cambridge, MA: Harvard University Press. Original edition, 1907.

James, W. 1975. *The Meaning of Truth.* The Works of William James. Edited by F. Burkhardt, F. Bowers and I. K. Skrupskelis. Cambridge, MA: Harvard University Press. Original edition, 1909.

James, W. 1978. *Essays in Philosophy.* The Works of William James. Edited by F. Burkhardt, F. Bowers and I. K. Skrupskelis. Cambridge, MA: Harvard University Press.

James, W. 1979. *The Will to Believe and Other Essays in Popular Philosophy.* The Works of William James. Edited by F. Burkhardt, F. Bowers and I. K. Skrupskelis. Cambridge, MA: Harvard University Press. Original edition, 1897.

James, W. 1981. *The Principles of Psychology.* 2 vols. The Works of William James. Edited by F. Burkhardt, F. Bowers and I. K. Skrupskelis. Cambridge, MA: Harvard University Press. Original edition, 1890.

James, W. 1982. *Essays in Religion and Morality.* The Works of William James. Edited by F. Burkhardt, F. Bowers and I. K. Skrupskelis. Cambridge, MA: Harvard University Press.

James, W. 1983. *Talks to Teachers on Psychology: And to Students on Some of Life's Ideals.* Edited by F. Burkhardt, F. Bowers and I. K. Skrupskelis. Cambridge, MA: Harvard University Press. Original edition, 1899.

James, W. 1985. *The Varieties of Religious Experience.* The Works of William James. Edited by F. Burkhardt, F. Bowers and I. K. Skrupskelis. Cambridge, MA: Harvard University Press. Original edition, 1902.

James, W. 1987. *Essays, Comments, and Reviews.* The Works of William James. Edited by F. Burkhardt, F. Bowers and I. K. Skrupskelis. Cambridge, MA: Harvard University Press.

James, W. 1988. *Manuscripts, Essays and Notes.* The Works of William James. Edited by F. Burkhardt, F. Bowers and I. K. Skrupskelis. Cambridge, MA: Harvard University Press.

James, W. 1992. *The Correspondence of William James.* 12 vols. Edited by I. K. Skrupskelis and E. M. Berkeley. Charlottesville: University of Virginia Press.

Johnson, D. R. 2010. *Nietzsche's Anti-Darwinism.* New York: Cambridge University Press.

Joly, H. 1883. *Psychologie des grands hommes.* Paris: Hachette.

Karakas, T. 2014. *Nietzsche et William James: Réformer la philosophie.* Paris: L'Harmattan.

Lange, F. A. 1866. *Geschichte des Materialismus und Kritik seiner Bedeutung in der Gegenwart.* Iserlohn, Germany: J. Baedeker.

Mencken, H. L. 1908. *The Philosophy of Friedrich Nietzsche.* London: T. Fisher Unwin.

Moore, G. 2002a. Nietzsche, Spencer, and the Ethics of Evolution. *Journal of Nietzsche Studies* 23, no. 1: 1–20.

Moore, G. 2002b. *Nietzsche, Biology, and Metaphor.* New York: Cambridge University Press.

Müller-Lauter, W. 1999. *Nietzsche: His Philosophy of Contradictions and the Contradictions of His Philosophy*. Translated by D. J. Parent. Urbana, IL: University of Illinois Press. Original edition, 1971.

Nietzsche, F. W. 1891. *Jenseits von Gut und Böse*. 2nd edn. Leipzig: Naumann.

Nietzsche, F. W. 1894. *Zur Genealogie der Moral*. 2nd edn. Leipzig: Naumann.

Nietzsche, F. W. 1954a. *Twilight of the Idols*. In *The Portable Nietzsche*, 463–563. Translated by W. Kaufmann. New York: Penguin Books. Original edition, 1888.

Nietzsche, F. W. 1954b. *Thus Spoke Zarathustra: A Book for All and None*. In *The Portable Nietzsche*, 463–563. Translated by W. Kaufmann. New York: Penguin Books. Original edition (serial), 1883–1891.

Nietzsche, F. W. 1966. *Beyond Good and Evil*. 2nd edn. Translated by W. Kaufmann. New York: Vintage Books. Original edition, 1886.

Nietzsche, F. W. 1968a. *On the Genealogy of Morals*. 2nd edn. In *The Basic Writings of Nietzsche*, 437–599. Translated by W. Kaufmann. New York: Modern Library. Original edition, 1887.

Nietzsche, F. W. 1968b. *The Birth of Tragedy*. 2nd edn. In *Basic Writings of Nietzsche*, 15–144. Translated by W. Kaufmann. New York: Modern Library. Original edition, 1872.

Nietzsche, F. W. 1968c. *The Will to Power*. Translated by W. Kaufmann. New York: Vintage Books.

Nietzsche, F. W. 1974. *The Gay Science*. 2nd edn. Translated by W. Kaufmann. New York: Vintage Books. Original edition, 1882.

Nordau, M. S. 1895. *Degeneration*. 2nd edn. London: William Heinemann.

Pearce, T. 2010. From 'Circumstances' to 'Environment': Herbert Spencer and the Origins of the Idea of Organism-Environment Interaction. *Studies in History and Philosophy of Biological and Biomedical Sciences* 41, no. 3: 241–252.

Perry, R. B. 1935. *The Thought and Character of William James*. 2 vols. Boston, MA: Little, Brown, and Company.

Pütz, M. 1995. *Nietzsche in American Literature and Thought*. New York: Camden House.

Ratner-Rosenhagen, J. 2011. *American Nietzsche: A History of an Icon and His Ideas*. Chicago, IL: University Of Chicago Press.

Richards, R. J. 1987. *Darwin and the Emergence of Evolutionary Theories of Mind and Behavior*. Chicago, IL: University of Chicago Press.

Richardson, J. 2004. *Nietzsche's New Darwinism*. Oxford: Oxford University Press.

Rolph, W. 1882. *Biologische Probleme: Zugleich als Versuch einer rationellen Ethik*. Leipzig, Germany: Engelmann.

Rorty, R. 1982. Nineteenth-Century Idealism and Twentieth-Century Textualism. In *Consequences of Pragmatism*, 139–159. Minneapolis: University of Minnesota Press.

Rorty, R. 2007. Pragmatism as Romantic Polytheism. In *Philosophy as Cultural Politics*, 27–41. Cambridge: Cambridge University Press.

Roux, W. 1881. *Der Kampf der Theile im Organismus: Ein Beitrag zur vervollständigung der mechanischen Zweckmässigkeitslehre*. Leipzig, Germany: W. Engelmann.

Salter, W. M. 1917. *Nietzsche the Thinker: A Study*. New York: H. Holt and Company.

Schopenhauer, A. 1844. *Die Welt als Wille und Vorstellung*. 2 vols. 2nd edn. Leipzig, Germany: F. A. Brodhaus.

Skrupskelis, I. K. 2007. Evolution and Pragmatism: An Unpublished Letter of William James. *Transactions of the Charles S. Peirce Society* 43, no. 4: 745–752.

Spencer, H. 1879. *The Data of Ethics*. London: Williams and Norgate.

Stack, G. J. 1992. *Nietzsche and Emerson: An Elective Affinity.* Athens, OH: Ohio University Press.

Taylor, E. 1990. William James on Darwin: An Evolutionary Theory of Consciousness. *Annals of the New York Academy of Sciences* 602, no. 1: 7–34.

Williams, B. 2004. *Truth and Truthfulness: An Essay in Genealogy.* Princeton, NJ: Princeton University Press.

5 Higher-order individuals
Truth and reality as organic systems

The guiding metaphor of nineteenth-century German idealism was reality-as-organism.[1] This metaphor challenges the mechanistic determinism of the Enlightenment by building a teleological principle of growth and maturation into history. For the idealist, development is not only real but is the optic through which everything must be viewed. Truth is in the whole, which is a self-realizing whole that expresses itself in history. Although the idealism of the early nineteenth century is not considered particularly scientific, it did challenge the same steady-state models of the world that were undermined by the scientific evolutionism that took hold by the century's end. Idealism thus prepared the ground for later evolutionism by forcing many thinkers to take development seriously. In Nietzsche's words, 'the minds of Europe were preformed for the last great scientific movement, Darwinism – for without Hegel there could have been no Darwin'.[2]

Although idealism waned in Germany starting in the 1830s, it was the reigning philosophy in the anglophone world during the height of the Darwinian controversies at the century's end. During William James's time, anglophone idealists like T. H. Green and F. H. Bradley at Oxford, and Josiah Royce at Harvard were all portraying reality as a self-developing rational whole. The dominant anglophone philosophies of the early twentieth century – the analytic logicism of Bertrand Russell and G. E. Moore in England and the pragmatism of William James and John Dewey in the US – were both forged in reaction to the metaphysical excesses of such views. Unlike the seminal analytic philosophers, however, the pragmatists reacted against idealism by giving a central role to reconstructed historicist and evolutionary concepts in their philosophies. The pragmatists were salvaging something of the idealist vision, rather than pivoting from a philosophy of development to a philosophy of static logic.

The present chapter traces William James's particular reaction to idealistic monism and to the notion of truth or reality as an organic system. Reality for James can be likened to a developing individual. However, James denies that reality comprises a totalized system or that synoptic knowledge is possible even in principle. Reality may be an organism, but it is an organism in the same sense in which we are organisms: finite, growing and fringed by an inchoate 'more'. At his most speculative, James posits an increasingly inclusive hierarchy of consciousnesses of which we are all a part. The highest consciousness here may be equated with God, but it is a finite God that is constituted in history by myriad sentient beings.

In undertaking this examination, this chapter situates the major concepts of James's later philosophy – pragmatism, pluralism and radical empiricism – in relation to his conception of evolution and individuality. James's conception of the individual, heretofore understood primarily as an individual human being, shows up in different ways at different levels of analysis. An examination of these levels allows James's ethics of self-transformation to appear in its broadest logical and metaphysical terms.

The most inclusive realizable whole

James's ethical vision is cast in overtly historicist terms in his 1891 essay 'The Moral Philosopher and the Moral Life'. This is James's only essay on ethical theory, in that it is the only publication where he defines goodness ontologically and pits traditional ethical theories against each other. According to James, goodness consists in the satisfaction of the interests or 'demands' of sentient beings. Goodness does not exist absolutely but is always relative to demands. A world with no sentient beings would thus have no moral stakes, and a world with a single sentient being would have its moral significance defined wholly in relation to that being. No simplistic Benthamite, James defines 'demand' in an intentionally open-ended way. A Jamesian demand may be for literally anything under the sun. Since no demand is absolutely disallowed, a demand can only be called into question by being incompatible with other demands. Given this meta-ethical framework, James concludes that our only 'categorical imperative' is to work toward a cooperative universe that allows for the greatest possible satisfaction of mutually compatible demands. This is the ideal of the *most inclusive realizable moral whole*, which James attributes in his lecture notes to his idealist colleague Josiah Royce.[3] This is not a purely negative ethics but a positive one of fostering and maximizing each other's interests. James thus exhorts his readers to '[i]nvent some manner of realizing your own ideals which will also satisfy the alien demands – that and only that is the path of peace!'[4]

James is more than a mildly heretical utilitarian. Indeed, 'The Moral Philosopher' is not so much a contribution to traditional ethical debates as an attempt to radically reconstruct them. James is upfront about his aim, opening the essay as follows:

> The main purpose of this paper is to show that there is no such thing possible as an ethical philosophy dogmatically made up in advance. We all help to determine the content of ethical philosophy so far as we contribute to the race's moral life. In other words, there can be no final truth in ethics any more than in physics.[5]

James treats ethical ideals in the same fashion as he treats other beliefs: Spawned from idiosyncratic conditions, they can be judged only in terms of their consequences. These consequences include their work in constructing the social and natural environment in which they operate – a kind of moral niche construction that alters the conditions for future ideals. There is thus an ineliminable historical dimension to inquiry that precludes the finalizing of norms at any stage. What we do now affects the very meaning and content of the good.

'The Moral Philosopher' ends by giving this historicist idea a religious cast. According to James, the most inclusive realizable moral whole is possible

> only in a world where there is a divine thinker with all-enveloping demands. If such a thinker existed, his way of subordinating the demands to one another would be the finally valid casuistic scale; his claims would be the most appealing; his ideal universe would be the most inclusive realizable whole ... therefore, we, as would-be philosophers, must postulate a divine thinker, and pray for the victory of the religious cause.[6]

Such a conclusion is surprising in an essay that begins by discarding a priori speculation in favour of a radically inductive approach to ethics. However, James is not positing a resting place to which history will be one day be delivered (by, say, Spencer's progress or Hegel's Absolute Spirit). Rather, a divine thinker for James represents a regulative ideal, or a hypothetical point at which the best moral equilibrium will have been reached. It is a practical posit that provides an inchoate goal to orient oneself by and strive toward. The idea of a fully realized maximally inclusive moral whole is supposed to energize us to move toward that state today, even if such movement is necessarily open to correction.

Such a posit is valuable to James because it unleashes what he calls 'the strenuous mood':

> The deepest life, practically, in the moral life of man is the difference between the easy-going and the strenuous mood. When in the easy-going mood the shrinking from present ill is our ruling consideration. The strenuous mood, on the contrary, makes us quite indifferent to present ill, if only the greater ideal be attained.[7]

There is no better selective environment for ethical ideals than a strenuous mood. What is repugnant is no longer tolerated, and what is valued comes into focus. To employ the Spinozan parlance of *Talks to Teachers*, James is promoting energetic moral agency *sub specie boni* (under the aspect of the good), as opposed to shrinking from challenges *sub specie mali* (under the aspect of the bad). The strenuous mood thus increases the pressure of the 'struggle for existence' among ideals – which for James means a *struggle among ideals*, not (with the social Darwinists) *brutal struggle as an ideal*.

'The Moral Philosopher' extends the moral psychology of *The Principles of Psychology* to the historical level. In mediating their habits, individuals also mediate society's 'habits' – a process that is most effective when one is aiming to constitute moral relations in society as a mirror of the moral preferences of a hypothesized divine being. James had already argued in the *Principles* that the highest act for the individual is the resistance of an embodied habit for the purpose of some ideal end. Similarly, James argues in 'The Moral Philosopher' that historical moral development proceeds through the breaking of rigid societal 'habits', such that 'the *highest* ethical life ... consists at all times in the breaking of rules which have grown too narrow for the actual case'.[8] In this way 'The Moral Philosopher' resembles James's earlier essay 'Great Men, Great Thoughts, and the Environment' (1880), which construes socio-historical change on the model of ontogeny.[9] 'The Moral Philosopher' simply emphasizes the moral dimension of this process. Whether in the moral development of individuals or of history writ large, the developing system is subject to non-directed variation that has unpredictable and non-predetermined effects.

James's call for strenuous activity in the name of a posited hypothetical divinity might sound naïve: Countless atrocities have been committed with great vigour in the name of some religion or other. However, religion in the Jamesian sense has little to do with churches or other institutions, and it is actively opposed to rigid dogmatism. It is a matter of one's general orientation toward existence, especially as this affects the project of achieving the most inclusively demand-satisfying intra-psychic and societal configurations. This may involve postulating a spiritual dimension in experience, but it is a spiritual dimension whose significance derives from its relations to communally shared moral gains in real historical time.

Truth

As in his ethical essays 'The Moral Philosopher and the Moral Life' and 'The Will to Believe', James is not simply taking a side within an extant debate. He is offering a historicist vision of an evolving world of which truth is one aspect. Indeed, given that truth for James is understood in terms of practical effects, his historicist theory of truth may properly be subsumed by his historicist moral philosophy. 'Truth' is a name for a certain type of good, that is, a useful functional relation between beliefs and their consequences within the developing texture of experience. The shape of this truth changes with the shape of the systems in which it inheres. This is how truth functions within James's philosophy of pragmatism.

Pragmatism

James introduced his pragmatism to the world in an 1898 lecture in Berkeley, California titled 'Philosophical Conceptions and Practical Results',[10] outlined his vision in *Pragmatism* (1907) and refined it in *The Meaning of Truth* (1909). James's theory is based upon an adaptation of key ideas from his old friend Charles S. Peirce. Of particular importance is Peirce's doctrine in 'How to Make Our Ideas Clear' (1878) that the meaning of a concept consists in its practical effects. This is Peirce's 'pragmatic maxim': 'Consider what effects, that might conceivably have practical bearings, we conceive the object of our conception to have. Then, our conception of these effects is the whole of our conception of the object'.[11] For instance, to call an object 'hard' is to say that certain effects – such as relative impenetrability – will obtain when the object is handled in a certain way. Anything other than such 'practical bearings' should not be admitted into the meaning of the object. To provide a more traditional philosophical example, Peirce's maxim could be applied like a solvent to the scholastic concept of substance: Substance by definition lacks sensory effects that are distinct from those of its properties (accidents) and is in this sense arguably bankrupt.

The idea of 'practical bearings' in Peirce should not be interpreted too broadly. Peirce is concerned with sensory effects and their relation to thought. In this sense he is doing traditional epistemology. In contrast, James develops a 'principle of pragmatism' that explicitly includes changes to *dispositions for action* among the relevant practical effects.[12] Meaning for James is not just an issue between the senses and the intellect. It is also between the intellect and the actions it guides. In terms of the physiological framework outlined in Chapter 3, it could be said that James refuses a theory of meaning that traverses only one half of the reflex arc. James completes the arc by connecting the sensory-cognitive half with the cognitive-behavioural half. Following John Dewey, moreover, the reflex 'arc' should really be understood as a *circuit*.[13] This means that 'downstream' elements are always-already influencing 'upstream' ones, such that sensation is never a wholly disoriented starting point but rather sensation-within-an-already-cognized-and-practically-engaged-situation. James thus enlarges Peirce's pragmatic maxim by understanding meaning to suffuse the entire sensorimotor dialectic in which the purposive individual is constituted as such in coordination with a lived world.

James's principle of pragmatism grounds a method for analysing and comparing concepts or theories. According to James, if two theories have identical practical results – if there is not *a difference that makes a difference* – then these theories are practically the same and debating them is a waste of time:

> There can *be* no difference anywhere that doesn't *make* a difference elsewhere – no difference in abstract truth that doesn't express itself in a difference in concrete fact and in conduct consequent upon that fact, imposed on somebody, somehow, somewhere and somewhen. The whole function of philosophy ought to be to find out what definite difference it will make to you and me, at definite instants of our life, if this world-formula or that world-formula be the true one.[14]

James gives the example of the chemist Ostwald, who denied any meaning to the difference between two competing explanations of the constitution of *tautomers* – a kind of isomer of organic compounds – because he could see no difference in experimental fact that could be made by one theory being true rather than the other. James claims that pragmatism here simply represents the attitude of empiricism, properly carried out.[15] Only empirically grounded concepts matter, and theories will be cashed out in terms of empirically observable effects.

James is thus trying to narrow philosophical discourse, or to separate the philosophical wheat from the chaff. In this way, James's pragmatic method works something like the twentieth-century logical positivists' verifiability principle. This is a very limited comparison, however. James's concept of empiricism (or experience) is broader than the positivist one, just as his concept of a practical effect is broader than Peirce's. James is also seeking to expand the universe of philosophical discourse. James wants to include *only* practical effects as meaningful, but he wants to include *all* such effects. The parts of philosophy that James finds suspicious are not its metaphysical speculation or moralizing, but its excessive logomachy and abstraction. As he tells a critical interlocutor in 1904, 'Your bogey is superstition; my bogey is desiccation'.[16]

Humanism

James's pragmatic theory of meaning is the basis of his original theory of truth. Commonly known as the 'pragmatist theory of truth', James typically follows his British supporter F. C. S. Schiller in calling it 'humanism'. The crux of James's position is that a belief is true to the extent that it 'works' or is 'useful' or 'satisfactory'. James's view is subject to misunderstandings, in part because of his folksy and popular presentations of his concepts. To be clear, James is not equating the truth of a belief with its contribution to our Darwinian fitness – a caricature of pragmatism to which no one subscribes. Rather, James's view is that 'true is the name of whatever proves itself to be good in the way of the belief', where survival is one optional interest among others.[17] Moreover, this does not amount to the commonsense claim that truths are useful because they are true; James is not pointing out that knowing things comes in handy. He is saying that a belief is useful *if and only if* it is true. That is, James is offering 'truth' as an umbrella term for those beliefs that prove themselves useful in the course of experience.[18]

Like 'The Will to Believe', James's humanism is easily parodied as promoting a kind of wishful thinking. James's position might seem to be that one can and should believe whatever one desires. Such a construal is unfair because James restricts beliefs in a coherentist manner. A belief cannot be true if it too baldly contradicts one's existing body of beliefs: 'True ideas are those we can assimilate, validate, corroborate and verify. False ideas are those we cannot'.[19] James here relies upon the inherently conservative nature of belief systems, whether at the social or individual level. Radically novel beliefs are rarely given credence. Indeed, there is often the opposite need to open up belief systems rather than closing them down. This is not typically achieved by radical jolts but by building connective tissue. In James's words, 'New truth is always a go-between, a smoother-over of transitions. It marries old opinion to new fact so as ever to show a minimum of jolt, a maximum of continuity'.[20] Anyone who thinks the pragmatist can believe whatever sounds pleasant should try this and see how it works. Believing that one can fly is an extremely difficult feat, for someone of any philosophical persuasion. This belief also does not lead to flight for the pragmatist any more than it does for anyone else. This is a recalcitrant practical effect that has been corroborated inductively for millennia.

The charge of wishful thinking also does not recognize the depth of James's fallibilism and historicism. James always takes the long view. To judge the truth of a belief is to contextualize it inductively within the grand sweep of a never-foreclosed process of verification. The results are never fully in. The problem is not just that we do not yet know what is already true – as if we are getting better and better at polishing a glass through which we are viewing a static extant truth. More radically, it is that truth is still being constructed. In today's epistemology, it could be said that James does not distinguish between truth and justification, where the former is a static relation and the latter a concrete process. Truth for James is a relation that is traced out in time and obtains by degrees. Quite literally, 'Truth happens to an idea. It becomes true, is made true by events. Its verity *is* in fact an event, a process: the process namely of its verifying itself'.[21] James thus agrees with the Hegelian that not even truth escapes time and change. Unlike the Hegelian, however, James does not posit that the self-realization of truth is implicit in reality or determined by an absolute rationality. James places us on shifting sands: 'it means a real change of heart, a break with absolutistic hopes, when one takes up this inductive view of the conditions of belief'.[22]

James's definition of truth as satisfactoriness immediately raises the question of how to define 'satisfactoriness'. However, James contends that this idea will 'admit of no definition, so many are the ways in which these requirements can practically be worked out'.[23] Truth has indefinitely many facets, corresponding to the indefinitely many ways in which a belief may prove itself meaningful by way of its practical effects. This means that novel forms of truth may emerge over time. Indeed, we currently engage in modes of thinking and acting – including entire symbolic systems, theoretical frameworks and modes of social interaction – that were literally unthinkable at earlier points in history. For James, such historically emergent modulations of human existence are literally modulations of truth.[24] James therefore does not offer a new explanatory account of truth but a kind of existential orientation: We will actively participate in the coalescence of truth as we obtain increasingly expansive vantage points from which to survey it inductively. A more expansive view will make us wiser but will never license a dogmatic curtailment of open questions or of potential new modes of existence. The open-endedness of James's position is radical.

This is not what most epistemologists are looking for in an account of truth. It seems like a bait-and-switch to claim to be offering a theory of truth and then to claim that this matter cannot be finally settled because it is subject to ongoing evolution. There is thus an impulse to domesticate James by assimilating him to more traditional views. H. S. Thayer, for example, claims that James admits the existence of useless truths but is only interested in analysing the pragmatic meaning of useful ones. This means that James is not redefining truth in terms of practical effects but merely shining a spotlight on certain kinds of truth.[25] Such a reading makes James appear more respectable. This is a waste of a radical philosopher. It is more instructive to let James stand as a clear example of someone who supplants all rounded-out explanatory accounts of truth with a wholly inductive and descriptive account. Let us make use of our intellectual outliers.

Meaning and function

James does not characterize his humanism in these terms because he is a careless philosophical dilettante. He does so out of a principled belief in the indeterminacy of function, which is grounded in his Pragmatic Image of Darwinism. James addresses the indeterminacy of function already in his first essay on Herbert Spencer. Here James denies Spencer's contention that mind *just is for* corresponding to the environment. According to James,

> Spencer's formula has crumbled into utter worthlessness in our hands, and we have nothing to replace it by except our several individual hypotheses, convictions, and beliefs. Far from being vouched for by the past, these are verified only by the future. They are all, in some sense, laws of the ideal. They have to keep house together, and the weakest goes to the wall. The survivors constitute the right way of thinking.[26]

To use Spencer's own phrase, James is advocating for a kind of 'survival of the fittest' among beliefs. This is consonant with James's humanism, where truths are those beliefs that prove themselves in experience. Notably, however, James here includes beliefs about the very function of the mind that generates beliefs. The function of mind is not determined by its past uses or by a pre-given cosmic teleology. What mind is for is partially up to us.

This is a challenge to any purely backward-looking account of function. One common way of understanding function in today's philosophy of science is *etiological*: A trait's function is explained in terms of its evolutionary history, typically construed in selectionist terms.[27] The heart makes sounds and has a certain colour and odour, but the pumping blood in particular explains its perseverance throughout evolutionary history. The pumping of blood may therefore be singled out as the organ's function. This view is intuitive, but it is challenging to employ in practice because of the inherently complex nature of evolution: Holding a piece of morphology constant, a shifting environment may change its function or render it useless; holding the environment constant, variation may introduce new functions or co-optations of old functions. The co-opting of an extant trait for a new purpose has been termed 'exaptation'.[28] The feathers of birds, for instance, seem to have been used for thermal regulation before being exapted for flight. Exaptation may be enabled by genetic mutation, or it may result simply from individuals learning new uses of extant structures during ontogeny and spreading this knowledge through imitation.

As a result, even if it were possible to determine what series of roles some piece of morphology has played in the past, this does not show what it might do in the future. The same may be said of social structures or bodies of belief. James's contemporary Nietzsche agues this point in his analysis of the function of punishment in society:

> there is for historiography of any kind no more important proposition than the one it took such effort to establish but which really *ought to be* established now: the cause of the origin of a thing and its eventual utility, its actual employment and place in a system of purposes, lie worlds apart; whatever exists, having somehow come into being, is again and again reinterpreted to new ends, taken over, transformed, and redirected by some power superior to it; ... and the entire history of a 'thing', an organ, a custom can in this way be a continuous sign-chain of ever new interpretations and adaptations whose causes do not even have to be related to one another but, on the contrary, in some cases succeed and alternate with one another in a purely chance fashion.[29]

James makes this point as well, by defining truth in terms of the various meanings that may accrue to a belief over time. The indeterminacy of meaning is underscored in James's future-oriented stance toward novel ideas, which are prized as raw material for individual or social development. Variation as such is valuable, because the concrete meaning of any given idea or action is always partially unknown. In Charlene Seigfried's words, 'Darwin emphasizes the fact that differentiations of all kind, even seemingly nonfunctional ones like the many beautiful colors of flowers, contribute to survival, while James emphasizes the corollary that this adaptive value is never known beforehand'.[30] Past function does not legislate present function, although it may constrain or prefigure it.

Notably, function may also be viewed synchronically rather than retrospectively, in terms of the various systems in which a trait now plays a role.[31] This *causal* view is much more promiscuous than the etiological view. The etiological view may face epistemic difficulties, but it is at least seeking a simple answer. In contrast, the causal view may define the function of a trait (or belief) in terms of indefinitely many kinds of effects in indefinitely many different (organic, psychological, social, epistemological, cosmic) systems. The etiological view cuts a particular swath through a vast thicket of roles played, in order to arrive at one particular sense of what a trait is 'for'. As a selective abstraction, it has limited uses in particular contexts. It does not license normativity in a strong sense, nor is it the only source of normativity.

If James emphasizes the open-endedness or indeterminacy of truth, he nevertheless insists upon its inherently constrained nature. He makes this point, too, in evolutionary terms:

> The most primitive ways of thinking may not yet be wholly expunged. Like our five fingers, our ear-bones, our rudimentary caudal appendage, or our other 'vestigial' peculiarities, they may remain as indelible tokens of events in our race-history.... You may alter your house *ad libitum*, but the ground-plan

of the first architect persists – you can make great changes, but you cannot change a Gothic church into a Doric temple.[32]

James's talk of an evolutionary 'ground-plan' resembles biologists' talk of '*Baupläne*' that are common to members of high-level taxonomic groups, and his architectural imagery prefigures Steven Jay Gould and Richard Lewontin's argument that certain features of organisms are like *spandrels* (by-products of archways) in that they exist only for holistic structural reasons.[33]

James's biological metaphors are interesting here, given that he had already provided a *literally* biological account of cognitive structures and their attendant 'necessary truths' in *The Principles of Psychology*. In his humanism, James is talking less about genetically inherited cognitive structure and more about the social evolution of commonly held truths. It is unclear the extent to which the social-evolutionary account of truth in *Pragmatism* is meant to supplement or replace the literally Darwinian account in the *Principles*. The two processes must conspire in a manner that is difficult to tease apart.

Absolute truth

Given all of this talk of indeterminacy, it may be surprising to learn that in *Pragmatism* and *The Meaning of Truth* James posits something called 'absolute truth'. This idea bears the stamp of Peirce's hoary notion of the 'end of inquiry' in which all questions will have been agreed upon by the community. According to James, 'The "absolutely" true, meaning what no farther experience will ever alter, is that ideal vanishing point toward which we imagine that all our temporary truths will some day converge'. James immediately places this concept in pragmatic context:

> This regulative notion of a potential better truth to be established later, possibly to be established some day absolutely, and having powers of retroactive legislation, turns its face, like all pragmatist notions, toward concreteness of fact, and toward the future. Like the half-truths, the absolute truth will have to be *made*.[34]

The 'intellectualist' posits a kind of static and logical truth to which we putatively already have access, although there is no definite indication whether one is standing in this relation at any given time. Thinking such a view vacuous and unverifiable, James defines truth in terms of the historical accretion of all manner of inductive and concretely experienced truths. James thus does not deny truth but instead challenges 'the pretence on anyone's part to have found for certain at any given moment what the shape of that truth is'.[35]

This position maps onto James's historicist ethical philosophy in 'The Moral Philosopher and the Moral Life', where the most inclusive moral whole should be posited as a regulative ideal. Because James's humanism is an aspect of his practical philosophy, the coalescence of absolute truth would be coordinate with the coalescence of ethical ideals represented by the most inclusive realizable moral whole. 'The Moral Philosopher' and James's humanism therefore outline the same philosophical territory with different emphases. This is the hypothetical future point toward which we should strive with all of the resources of the strenuous mood. This goal may never be attained. Indeed, the very idea of absolute truth for the pragmatist is merely an induction based on the experience of some amount of coalescence of truths up to this point. Absolute truth is thus a self-verifying posit, for which no particular amount of verification is guaranteed.[36] James thus describes pragmatism as an embrace of 'the open air and possibilities of nature, as against dogma, artificiality, and the pretence of finality in truth'.[37]

Even if one is prepared to accept James's basic concept of absolute truth as a practical posit, it is reasonable to ask for further detail about what constitutes communally shared truth on a pragmatist worldview. James makes grand statements about the sweep of history, but what can be said more concretely about objectivity? This is an important topic that will not be addressed at length in this space. It is worth noting, however, that among the most pragmatist-friendly reconstructions of objectivity in recent decades have been by feminists, especially those who are concerned with the concrete contexts in which knowledge is generated.[38] Some of these writers work within the history and philosophy of science, and a few take up the pragmatist tradition explicitly.[39]

A starting point among feminists is typically the recognition that objectivity is an ongoing negotiation among various types of demand, rather than something to be posited in a facile gesture as if to ward off the real work of thinking in the world. For instance, Helen Longino outlines a feminist conception of objectivity that is broadly consonant with James's pragmatism. Longino provides six criteria for objectivity, which have no inherent order of priority:[40]

- Empirical adequacy
- Novelty
- Ontological heterogeneity
- Complexity of relationship
- Applicability to current human needs
- Diffusion of power

A common mistake, according to both Longino and James, is to grant a special status to the first criterion, understood as a fit between a model or theory with observed data. There is a tendency to view empirical adequacy as constitutive of objectivity whereas other factors are merely contextual. A more robust concept of objectivity examines how empirical adequacy itself is constituted, while not presuming that it stands atop a hierarchy of demands. Indeed, the criteria given above are largely ways of dissolving hierarchies in specific contexts: To prize novelty is to value variation as a resource for the undermining of sub-optimal entrenched structures; and to prize ontological heterogeneity and complexity of relationship is to refuse to construe certain types of entity or causality as more real than others, or as a kind of Archimedean point from which causation issues absolutely.

These principles underlie James's moral and epistemic vision, which challenges structures by railing against simplistic accounts of developing systems that give a special metaphysical status to one particular direction of causation. Longino's criterion of applicability to human needs is also represented by James's pragmatic method, which interprets theoretical issues in terms of practical effects. This is what Dewey was getting at when he claimed that thinkers' real work consists in attending to the 'problems of men' – with the proviso that an anti-hierarchical philosophy of science may broaden this phrasing to include the problems of a panoply of non-ranked identities.

If the pragmatist or feminist will construe objectivity as constituted rather than given, the intellectualist will claim that such a conception could not possibly ground normativity. The Jamesian rejoinder to this criticism is that a wholly logical account of truth could only license an empty and stipulative kind of normativity. Truth is not normative simply because it is truth but because of all of the particular ways in which truths function concretely. Therefore, 'humanism explicates the summarizing word "ought" into a mass of pragmatic motives from the midst of which our critics think that truth itself takes flight'.[41] If this view licenses only an indeterminate and negotiable kind of normativity for beliefs (or organic functions), then perhaps this is because no other kind of normativity exists.

Reality

Like his humanism, James's metaphysics is usefully construed in organic terms. Indeed, to separate these two parts of James's thought is in some ways artificial. Truth for James is not a scorecard standing apart from the world and taking account of it. It is an evolving system that is one facet of an evolving reality. That is, each new truth is 'a positive addition to the previous reality',[42] which must now be negotiated along with the rest of experience.

James illustrates this idea with his snowball metaphor in *Pragmatism*:

> The case [of truth] is like a snowball's growth, due as it is to the distribution of the snow on the one hand, and to the successive pushes of the boys on the other, with these factors co-determining each other incessantly.[43]

The snowball (the accrued mass of accepted truths) maintains traction with its embedding reality (the terrain), while the boys (inquirers) codetermine its shape, size and direction. The boys' pushes are partially determined by the snowball's prior extant properties and velocity; the terrain is heterogeneous and changing; and the game is 'incessant' in that it has no natural stopping point. This is an open-ended evolutionary account of a world in which truth is understood through its concrete functions.

Interestingly, this metaphor resembles Susan Oyama's tent-raising metaphor in *The Ontogeny of Information* (1985), which is the seminal work of developmental systems theory in biology. According to Oyama, ontogeny can be likened to 'the idea of campers raising and stabilizing a tent pole by pulling in opposite directions'.[44] The point is that stability is a provisional dynamic accomplishment of various factors whose contributions only make sense relative to one another. Patterns of ordered change should not be attributed disproportionately to any one environmental or internal factor. Oyama especially critiques the concept of genetic information as a blueprint or code for ontogeny, as this view is preformationist and posits a special type of privileged cause. She therefore suggests replacing the preformationist concept of genetic information with the idea of information as emerging over time from contingent conditions within the developing system itself. In other words, *information has an ontogeny*. Truth for James is similarly a cooperative, dynamic and provisional accomplishment. Strikingly, James even uses the same phrase as Oyama – 'differences that make a difference' – to refer to factors that have a practical effect on the developing system.[45] On James's view, *truth has an ontogeny*, which is one aspect of *reality's ontogeny*.

Pluralism

In his introduction to his father's posthumous *Literary Remains* (1884), James claims that the difference between monism and pluralism is 'the deepest of all philosophic differences'.[46] James here intends to highlight a difference between his father and himself. Henry James, Sr. was a Swedenborgian mystic for whom God was the only truly active principle in the universe. According to the elder James, God's purpose is realized only insofar as humanity is cleansed of the illusion of selfhood. Although William appreciated his father's view that redemption occurs within history instead of beyond it, he could not countenance a metaphysics that denies the significance of individual life. In this respect, he stands closer to his father's friend Emerson, whose 'most characteristic note' was '[t]he faith that in a life at first hand there is something sacred'.[47] James therefore styles himself as a rebel, claiming that 'the most serious enemy' of his father's work 'will be the *philosophical* pluralist'.[48]

James repeats the claim that the question of monism and pluralism is the philosophically deepest one throughout his career. This includes the fourth lecture of *Pragmatism*; the aborted book manuscript 'The Many and the One',[49] which was to be his only technical exposition of his thought for a professional audience; and his unfinished final work *Some Problems of Philosophy*, which addresses a series of metaphysical questions in the form of an introductory philosophy text.[50] In the latter text, James focuses on what he calls 'noetic monism', or the absolute idealist's view that 'the world exists no otherwise than as the object of one infinitely knowing mind'.[51] The idealist view can be understood as making an analogy to individual consciousness: Just as we finite knowers experience a many-at-oneness that is knit together by relations of conjunction and disjunction by virtue of our rational consciousness, the world as a whole hangs together in its rational relatedness by virtue of an all-encompassing mind. Without an infinite mind, the idealist believes, the world would dissolve into an irrational plurality with no unity or tightness of fit among its parts.

As with any open metaphysical question, James believes that the question of monism and pluralism may legitimately be answered on practical grounds. Monism seems to make the world more rational and provides a sense of comfort, but it makes mysterious the specificity of finite consciousness and brute sensation; creates a facile theodicy since evil is now construed as part of a broader plan; and is fatalistic rather than allowing for genuine possibility. To this view James contrasts his pluralism, which holds that 'everything in the world might be known by somebody, yet not everything by the same knower, or in one single cognitive act'.[52] James's pluralism respects the bruteness of sensation and novelty, which is valued as raw material for a finite rationality that is produced by finite knowers. It also dissolves the problem of evil by claiming only that concrete evils (in the plural) may be reduced if we are willing to act against them. James's pluralism is in this way *melioristic*, that is, it is intended to cultivate the 'strenuous mood' so that individuals may alter a genuinely plastic world. In James's words, pluralism

> is neither optimistic nor pessimistic but *melioristic* rather. The world, it thinks, *may* be saved, *on the condition that its parts shall do their best*. But shipwreck in detail, or even on the whole, is among the open possibilities. There is thus a practical lack of balance about pluralism, which contrasts with monism's *peace of mind*.[53]

James's phrase 'practical lack of balance' is telling. Unlike the monist's view that the world is already completed – even if history has yet to fill in some of the details – pluralism represents something of the actual rhythm of life. This includes the sense of continually falling into or out of step with posited ideals; the culminations and frustrations of our projects; and the generative dialectical relationship between active striving and passive faith in circumstances, where neither one may replace the other absolutely. In many senses, the off-kilter nature of existence is what pulses it forward. Dialectical versions of idealism like Hegel's take such rhythms seriously, but instead of working this logic empirically they blow it up into a totalizing system that subsumes the concrete under monstrous abstractions.

James's pluralism generalizes deep lessons that he first learned in the context of Darwinism and physiology. For instance, James often uses the terms 'datum' and 'gift' to refer to certain realities that are accepted as given in one's theories. Every theory stops somewhere. Darwin's theory showed that accidental variation could be accepted as a gift in this sense. In his social evolutionism, James argues that the contributions of individuals to society are similarly gifts that cannot be explained wholly as resultants of social or geographical pressures. In pluralism, this lesson is generalized to the ultimate level. If organic or social variation is a gift that must be 'begged' by evolutionists, then 'being is a datum or gift and has to be begged by the philosopher'.[54] The only question is whether this datum must be begged all at once – say, in a 'big bang' event that sets up a deterministic universe – or in instalments, such that novelty flows into reality continually. James opts for the latter view. This is a deep refusal of the totalizing ambitions of either absolute idealism or other traditional religious and philosophical systems: We must accept a series of inscrutable ontological gifts; being may comprise a 'concatenated union' in which the parts hold together by way of intermediary connections, but not in a monistic 'total conflux' where every 'each' is nothing without the absolute 'all'.[55] James represents such a pluralistic universe with a rare *gustatory* metaphor for knowledge: 'Not unfortunately the universe is wild – game flavored as a hawk's wing'.[56]

Radical empiricism

James is like the absolute idealist in that he models reality on mind: The observed features of experience are taken to be indicative of the nature of reality in general. However, James denies the need for an infinite or totalizing mind. Here James argues a point that he had argued against Chauncey Wright in the Metaphysical Club in the 1870s,[57] which reappears in his 1884 essay 'On Some Omissions of Introspective Psychology' and in his famous chapter 'The Stream of Thought' in *The Principles of Psychology*:[58] Namely, experience is not a *train* of atomistic elements but a *stream* in which continuity is given. Connection and disconnection are coeval, in both consciousness and the world. Reality is *dynamically continuous*, with its own native kind of finite rationality.

James first describes his metaphysics as 'radical empiricism' in his 1897 preface to *The Will to Believe*. Here he identifies *empiricism* with fallibilism, and he claims that his view is *radical* in treating pluralism as its (fallible) metaphysical hypothesis.[59] If radical empiricism in 1897 is little more than pluralism, then by *The Meaning of Truth* (1909) James adds further postulates. Radical empiricism now includes the methodological principle that philosophers may only discuss matters drawn from experience, combined with the claim that both disjunctive and conjunctive relations are equally matters of direct experience.[60] Philosophy must be grounded in the concrete, but not in a falsely abstracted version of the concrete that alienates discrete parts from their context and posits them as pre-existent. Radical empiricism takes experience seriously, but in doing so finds that 'experience' or 'empirical' in their traditional sense have been intellectualistic abstractions.

A key finding of radical empiricism is that consciousness does not exist.[61] At least, this is James's provocative way of stating that *transcendental* consciousness does not exist. Experience does not require a Kantian metaphysical lynchpin to make it possible. Instead of a transcendental subject, James posits a level of 'pure experience' that is ontologically prior to both subject and object. The knower and the known are not basic features of reality but derivative functions of pure experience. James's claim that relations are as real as discrete things therefore works on multiple levels: It is true of any given known objects that the various relations among them are as real as they are; and it is true of the very epistemological subject-object relation that it does not pre-exist the knowledge relations in which it can be reflectively observed. James is directly contradicting the claim by some epistemologists that phenomenological analysis reveals experience always to be structured by a subject and an object. According to James, such a framework comes by way of a retrospective *addition*, not analytical subtraction:

> the usual view is, that by mental subtraction we can separate the two factors of experience.... Now my contention is exactly the reverse of this. *Experience, I believe, has no such inner duplicity; and the separation of it into consciousness and content comes, not by way of subtraction, but by way of addition* – the addition, to a given concrete piece of it, of other sets of experiences, in connection with which severally its use or function may be of two kinds.[62]

James thus claims that the self-same portion of experience may serve both as a knower in one functional context and a known in another. Knowledge is a matter of relations traced among parts of experience in which certain portions of experience know others but may in turn be known. There is in this system no stable point of absolute or transcendental knowing, whether for the individual or for reality itself. Knowledge is traced by functional relations within pure experience, which is connected to itself in various ways but is not propped up as a whole by anything. To the Kantian who claims that the 'I think' must underlie all experience, James responds with the amusingly corporeal suggestion that this is merely an illusion encouraged by our experience of our own breathing.[63] We are used to the idea that some life-sustaining process is always occurring to which our attention may be drawn, but this does not justify an ever-present unseen knower. To coin a phrase, Kant's transcendental unity of apperception may only be the *empirical continuity of respiration.*

Radical empiricism is not without its challenges. James considered the most recalcitrant problem for radical empiricism to be that of how two minds can know one thing. A danger is that radical empiricism may suggest a pluralistic solipsism on which self-enclosed subjects emerge from pure experience and do not share a public world.[64] James responds that on radical empiricism two subjects can know the same pen – for example, can see it from different angles – rather than two different private versions of a pen. Nevertheless, James cannot construe this pen as an object standing self-constituted in the world. The pen is primordially neither subjective nor objective. This mirrors how it is experienced: In the first place it appears just as a pen (with no qualification), and only reflection forces us to postulate two versions of it – one that belongs in a history of the external world and another that belongs in a history of our subjective experience. For James, the same pen may belong to both of these histories in the same sense that a single point may lie on two lines. There is some mystery in this view, but it provides a more directly realist epistemology than commonplace representationalism, which posits innumerable copies of the same objects and is saddled with the insoluble problem of accounting for the putatively mind-independent nature of objects that can be known only through some consciousness or other.

In his final years, James developed his pluralism and radical empiricism under the strong influence of Henri Bergson. James considered Bergson a great critic of the abuse of abstract concepts ('intellectualism') and a visionary regarding the evolution of reality. The following letter to Bergson expresses James's appreciation of his book *L'évolution créatrice* ('Creative Evolution'):

> To me at present the vital achievement of the book is that it inflicts an irrevocable death-wound upon Intellectualism. It can never resuscitate!... You will be receiving my own little book 'pragmatism' with this letter. How jejune and inconsiderable it seems in comparison with your great system! But is so congruent with parts of your system, fits so well into interstices thereof, that you will easily understand why I am so enthusiastic. I feel that at bottom we are fighting the same fight, you a commander, I in the ranks. The position we are rescuing is 'Tychism' and a really growing world. But whereas I have hitherto found no better way of defending Tychism than by affirming the spontaneous addition of *discrete* elements of being (or their subtraction), thereby playing the game with intellectualist weapons, you set things straight at a single stroke by your fundamental conception of the continuously creative nature of reality.[65]

This is an interesting shift in James's post-Darwinian worldview. James had long defended Peirce's doctrine of tychism, or the notion that chance is a genuine feature of reality. This allowed him to understand Darwinian variation as literally random, or at least as being underwritten by jolts of novelty that irrupt into the nervous system at a microscopic level. However, James now comes to believe that he had erred in understanding novelty to irrupt in discrete bits. James now agrees with Bergson (and with Peirce's concept of *synechism*) that nature is continuously creative. To posit discreteness among new elements of reality is to make cuts in pure experience at too deep a level. Discreteness comes with rational conceptual knowledge, which is a derivative feature of the world that is not present at the ultimate sources of variation. James makes this point in both Peircean and Bergsonian terms in an appendix to *A Pluralistic Universe*:

> if such a synechistic pluralism as Peirce, Bergson, and I believe in, be what really exists, every phenomenon of development, even the simplest, would prove equally rebellious to our science should the latter pretend to give us literally accurate instead of approximate, or statistically generalized, pictures of the development of reality.[66]

James thus ties his metaphysics explicitly to a statistical and fallible conception of science, which is just the kind of (Darwinian) science that he first drew upon to construe complex systems as nonlinear and subject to novel disruptions. Radical empiricism promotes fallible science and is a development of concepts from such science.

Radical empiricism leaves open further deep questions. Why does pure experience differentiate itself into subject and object at all? How many features of existence are properly treated as mysteries (or 'gifts') rather than as questions to be answered in conceptual terms? No metaphysical theory can answer every question that is put to it, nor is it guaranteed that all such questions will have answers. At some level, 'We may be in the universe as dogs and cats are in our libraries'.[67]

Panpsychism

James's pluralism and radical empiricism provide some sense of how he understands reality to have an ontogeny: Reality is structured by patterns of ordered change that are inflected by non-directed variations – differences that make a difference – and that in certain of their cognitive functions may be called 'knowledge'. There is something organic about this, but is it enough to warrant the analogy of reality to an *organism*?

As it happens, James does get more specific about this point. In particular, he postulates that reality may contain cascades of increasingly inclusive minds, such that each of our finite consciousnesses may be included within indefinitely many ascending levels of super-consciousness. Just as an individual's subconscious activities are somehow coordinate with a higher conscious life of which they have no inkling, each of our consciousnesses may similarly underlie higher levels of consciousness. James may not have openly argued this position as his own, but he was increasingly enthusiastic about it in the final decade of his life and defended its basic consonance with pluralism and radical empiricism. Especially when combined with his interests in psychical phenomena, mysticism and spiritualism, such a *pluralist panpsychism* gives us James at his most speculative.

In order to open the door to panpsychism, James had to reverse his earlier position on the 'compounding of consciousness'. In *The Principles of Psychology*, James had argued that simpler states of consciousness do not compound into more complex ones. Thus, for example, when one thinks of a series of objects singly, these various psychic states are not constituent parts of a more complex state in which one takes the objects as a collection. Instead, James argued that each state is a wholly new one, which has only a functional relation to the prior states. James wanted to avoid invoking transcendental machinery to synthesize a collection of representations. He also believed it was logical that an experience of a collection is something in addition to – and not comprehensive of – the experiences of the various objects collected. James does not give up on the idea that the latter view is 'logical'. Instead, he follows Bergson in arguing that logic is a limited tool that fails to penetrate reality. James was less committed to a particular kind of static and conceptual logic than he was to an intuitive understanding of reality as continuously developing and hierarchically inclusive in its levels of mentality: 'Logic being the lesser thing, the static incomplete abstraction, must succumb to reality, not reality to logic'.[68]

This shift allows James to describe an individual's mental states as aggregates of component states. It also allows him to posit that any given sentient being's consciousness is a component of a higher consciousness. James adapts such a conception from German psychophysicist Gustav Fechner and Polish philosopher Wincenty Lutoslawksi. James penned a preface to the former's *Little Book of Life after Death* (1904), as well as an introduction to the latter's *World of Souls* (1924).[69] James highlights in both places the idea of a hierarchy of souls that 'sum up' to a god-like figure that is enacted in time. The idea of history bodying forth divinity resonates with both absolute idealism and the writings of James's father. However, Lutoslawski and Fechner are closer to James's position that the divine is a finite product rather than a primordial creator. Indeed, this being is as much 'refractory' of its subordinate consciousnesses as it is generative or agential in its own right. This view avoids the problem of evil, as such a being lacks omnipotence and omniscience. As a result, however, it gives human beings serious work to do in eliminating evils in the plural.

This kind of panpsychism lends the open-ended diachronic theology of 'The Moral Philosopher and the Moral Life' a synchronic dimension. The point is no longer to posit a maximally inclusive hierarchy of ideals that *could* exist so that we may move toward this ideal; it is also to view the relationships among extant ideals within the hierarchy of consciousnesses *today* as constituting an imperfect divine being. We are already constituting God. In other words – like society in 'Great Men, Great Thoughts, and the Environment', ethical ideals in 'The Moral Philosopher and the Moral Life' and truth in *Pragmatism* and *The Meaning of Truth* – divinity has an ontogeny. As always for James, ontogeny here emphatically *does not* imply preformationism or asymmetric externalism. On the contrary, it implies non-directed variation that influences a contingent process of development in coordination with a relatively independent environment.

This last point is intended literally: God has an environment. According to James, one must

> be frankly pluralistic and assume that the superhuman consciousness, however vast it may be, has itself an external environment, and consequently is finite.... Yet because God is not the absolute, but is himself a part when the system is conceived pluralistically, his functions can be taken as not wholly dissimilar to those of the smaller parts – as similar to our functions consequently. Having an environment, being in time, and working out a history just like ourselves, he escapes from the foreignness from all that is human, of the static timeless perfect absolute.[70]

It is a striking claim that God has an environment, but this follows from the general pluralistic thesis that *everything* has an environment:

> Everything you can think of, however vast or inclusive, has on the pluralistic view a genuinely 'external' environment of some sort or amount. Things are 'with' one another in many ways, but nothing includes everything, or

dominates over everything. The word 'and' trails along after every sentence. Something always escapes. 'Ever not quite' has to be said of the best attempts made anywhere in the universe at attaining all-inclusiveness.[71]

As a matter of principle, any perspective taken can be situated relative to some 'more' that exceeds it. The monistic idealist posits an absolute being that, like the traditional Judeo-Christian deity, lacks any environment in relation to which it moves. It simply *is*, and expresses itself absolutely. This presents a great mystery: Why does the Absolute lapse into a world of appearance, evil and relative truth, where it then redeems itself through its own self-realization? On a pluralistic conception, the Absolute appears to be a reification of abstract properties taken to idealized limits that have no analogue in experience: an organism with no environment; a fully preformationist ontogeny; a force that issues absolutely instead of being relative to other forces; a synoptic perspective from which there is nothing new to be said; etc. Pluralism is more at home in the contingent messiness of the empirical world. This is what the universe looks like when it is modelled on actual minds.

Pluralistic panpsychism dovetails with a number of James's positions, which he understood to be mutually reinforcing. Already in *The Varieties of Religious Experience* (1902), for instance, James argued that unconscious activities may 'save' us by resolving our internal crises and giving us additional energies when we need them. James understands that his position can be given a purely secular reading. However, he speculates that there may be many finite gods (levels of super-consciousness) that influence human activities through the 'back door' of the unconscious. This idea is important to James because of its role in self-transformation, which in the *Varieties* is explained as a shift among 'centres of energy'.[72] If the unconscious is a permeable boundary to a wider spiritual realm, then this realm can innervate self-transformation and can in turn be affected by one's actions. This is an image of self-transformation as panpsychical niche construction. James believed that this spiritualistic interpretation of the unconscious was backed up by abnormal psychology and psychical research, two fields in which he actively participated.[73] As esoteric as all of this sounds, James maintains that the fruits of spiritual commerce are to be evaluated in terms of individual life. This grounds a pragmatist *soteriology*, or theory of salvation: 'when we touch our own upper limit and live in our own highest centre of energy, we may call ourselves saved'.[74]

Although this concept of salvation does not imply immortality, James does argue for the possibility of certain heterodox kind of immortality in an 1898 essay:[75] Namely, upon death one may recede back into an absorbing super-consciousness of which one had been a part. (James is explicitly pluralistic here: One is not absorbed into a monistic absolute, but rather into a finite super-consciousness among others.) James makes a loose case that one could be identical with the embedding super-consciousness, if it has been impressed with one's memories and in this sense meets the Lockean criterion for personal identity.[76] This is not the same as continuing on unchanged as the same individual, but it may be better: One would remain part of a growing world of environments and resistances, which is what gave life significance up to that point. There is something bloodless and spectral about the conception of immortality as being frozen in one's development state for immeasurably longer than one had ever been alive in the first place.

James's essay on immortality introduces a final fascinating speculation on the role of consciousness in individual and cosmic life. This is James's suggestion that consciousness may serve a *transmissive* function. Speaking as a physiologist, James points out that 'function' means only co-variation: One factor changes when another one does. According to James, we are free to choose among different metaphysical characterizations of how consciousness is a function of the brain. Does the brain *produce* consciousness as the liver produces bile, or does it *transmit* consciousness as a pipe organ transmits air? A transmissive view of consciousness would bring James's conception of mind closer to Kant's: Both portray our finite nature as negatively determining consciousness by imposing constraints or limits, in contrast to a hypothetical infinite (or relatively less determined) type of mind. For James, death here may be experienced as the lifting of a set of limits to reveal a new and different one, as when one is reabsorbed into an embedding super-consciousness.

In James's view, it is more parsimonious to view consciousness as a transmissive function of the brain. After all, this view does not require consciousness to be produced *de novo* at all times in each knower in the world. Instead, James follows Fechner in suggesting that consciousness and matter are coeval aspects of reality. Consciousness is already there, as the living side of matter. Direct spiritual communication is not always available, however. For unknown reasons, it occurs only with the shifting of a certain psychophysical threshold. One therefore experiences spiritual reality discontinuously, only at the wave-peaks of the rising and falling panpsychical mother-sea.

Conclusion

James was attempting to broker a novel position in the history of philosophy: a historicist empiricism, or a pluralistic historicism. In doing so, he employed the idealist metaphor of reality as developing organism. The world seems to hang together in a certain organic or developmental perspective. To analogize reality to an actual organic system, however, means understanding it as only *relatively* integrated, harmonized and well defined. Reality for James is an organism much as we are organisms: finite, developing and fringed by an inchoate 'more'. There is no totalizing perspective to be had. James thus preferred the term 'pluriverse' to 'universe' when pressed.

Another way to put this point is that James looks to the ontological middle. Meaning and truth, for instance, get produced literally in the *mediation* of cognitive and social variation for concrete ends. Lower levels of reality are interesting insofar as they feed such mediation, and absolute truth (or a spiritual dimension of existence) is interesting insofar as it influences the constructive activity that is happening today. To privilege an ontological basement (reductionism) or ceiling (absolute idealism) is a methodological presumption rather than an induction from experience. An alternative attitude is that middle-level phenomena – such as open-ended complex systems that exist always in an embedding environment – may be explained according to their own dynamic principles, which may extend as far in any ontological direction as we are able to project.

Truth in James's worldview is a label for beliefs that enter into satisfactory functional relationships within experience. There are as many ways for a belief to be true as there are ways of being functionally embedded (whether synchronically or diachronically) in an environment. Truths here are positive additions to reality rather than a distinct accounting of it. As relations between beliefs and their consequences accruing in time, truths are among the genuine growing realities that one must negotiate. Truth therefore has an ontogeny for James, just as information for Susan Oyama has an ontogeny: Rather than residing outside of the concrete processes that they guide, both truth and information are comprehensible *ways* in which these processes unfold. The world is an open-ended organic system marbled by a circulatory system of truth.

James's metaphysics of radical empiricism further shows the depth of his similarity with Oyama. The basis of Oyama's critique of information is her rejection of the ancient philosophical view that matter is intrinsically inert and therefore must be torqued into motion by some intentional superadded force. Without such a force, it would seem, neither the cosmos nor life could be set in motion. God has played this role in cosmogony. The informational gene has played this role in neo-Darwinism. The problematic dissolves, however, if matter is viewed as inherently dynamic. Similarly, James's radical empiricism is predicated upon a rejection of the classical empiricist view that disconnection is given and continuity must be superimposed. No transcendental machinery is necessary if reality is already dynamically continuous. Both James and Oyama thus eschew dubious metaphysical posits by refraining from reifying certain dichotomies – matter/form, connection/disconnection, relation/*relata* – and positing them as primordial features of the world.

For all of his criticism of intellectualist abstractions, there is something brazen in James's insistence that he is flying the banner of empiricism. Empiricism is associated by James with a 'tough-minded' attitude that stays close to facts and is often hostile to spirituality. Nevertheless, James claims this title even in his most speculative moments, as when he chides the Oxford idealists for being 'ignorant of the great empirical movement towards a pluralistic panpsychic view of the universe'.[77] James is co-opting – or *exapting* – the terms 'experience' and 'empiricism' for his purposes. James wants to take all of experience seriously, including those arational fringes that may lead into netherworlds. Such an attitude flows from a conception of meaning in terms of practical effects, combined with a conviction that metaphysical beliefs may have practical effects that a believer is free to embrace in the absence of disconfirming evidence.

This makes James less naturalistic or 'this-worldly' than, say, his fellow pragmatist John Dewey. Nevertheless, James's metaphysical views gain their significance in relation to concrete individual life – that is, in their work in co-constructing the individual's twofold habitual and social-natural environment. James's metaphysics is in this way grounded in an ethics of self-transformation, even when it posits a hierarchy of selves innervated by a spiritual realm. The image is of a pulsing creative hierarchical organism, the mediation of which is a continual source of moral significance and dramatic tension.

Notes

1 Mandelbaum 1971.
2 Nietzsche 1974/1882, §357.
3   The solution is by Royce's 'moral insight' – consider *every* good as a real good, and *keep as many as we can*. That act is the best act, which *makes for the best whole*, the best whole being that which prevails at least cost, in which the van-quished goods are least completely annulled

(ML, 185)

These notes are from James's only course on ethical theory, 'Philosophy 4: Recent Contributions to Theistic Ethics', which he taught once in 1888/1889.
4 WB, 155. Emphasis removed.
5 WB, 141.
6 WB, 161.
7 WB, 159–160.
8 WB, 158.
9 WB, 172.
10 P, 255–270.
11 Peirce 1878, 293.
12 P, 29. Peirce was quick to distance himself from this version of pragmatism.
13 Dewey 1896.
14 P, 30.
15 P, 31.
16 LWJ II, 208.
17 P, 42. Emphasis removed.
18 This provides an important *disanalogy* between James's humanism and Darwinism: In the latter theory, survival does not imply fitness. This would make 'the survival of the fittest' trivial. (It is not trivial, for example, if fitness is understood as a *propensity*.) On James's humanism, however, truth actually implies usefulness (and vice versa). On the tautology charge against natural selection, see Sober (1984).
19 P, 97.
20 P, 35.
21 P, 97.
22 MT, 40.
23 MT, 59–60.
24 Compare Williams (2004). Williams is critical of much traditional epistemology but distinguishes between truth and its historical manifestations (or 'truthfulness').
25 Thayer 1975. Thayer here draws upon a distinction between cognitive and pragmatic meaning found in Suckiel (1982).
26 EPH, 20.
27 Wright 1973; Millikan 1984.
28 Gould and Vrba 1982.
29 Nietzsche 1968/1887 II, §12.
30 Seigfried 1984, 267.
31 Cummins 1975.

32 P, 83.
33 Gould and Lewontin 1979.
34 P, 106–107.
35 MT, 143.
36 MT, 144.
37 P, 31.
38 Harding 1986; Haraway 1991.
39 Seigfried 1991; Lloyd 1995; Seigfried 1996; Sullivan 2001. Karen Barad (2007) puts forth a theory of 'agential realism' that resembles James's metaphysics of radical empiricism in that it makes subject and object derivative functions of a more primordial reality. Both Barad and James were trained as scientists and apply scientific concepts creatively in their philosophies. (Barad's view is inspired by the thesis of ontological indeterminacy in quantum mechanics.) See also *contextual objectivity* in Winther (in preparation).
40 Longino 1995, 386–389.
41 ERE, 129.
42 MT, 60.
43 P, 108.
44 Oyama 2000/1985, 27.
45 Oyama 2000/1985, 3. Oyama cites Gregory Bateson, not James, as her source for 'difference that makes a difference'. See Bateson (1972, 315).
46 ERM, 61.
47 ERM, 111.
48 ERM, 61.
49 MEN, 1–53.
50 James dedicates the latter text to Charles Renouvier, who in the early 1870s had swayed him toward both pluralism and a model of will as selective attention.
51 SPP, 71.
52 SPP, 68.
53 SPP, 73.
54 SPP, 74–75.
55 ERE, 52.
56 WB, 6. James is quoting obscure pamphleteer Benjamin Blood (1893). James was a champion of unconventional intellectual variation.
57 MEN, 150–154.
58 EPS, 142–167; PP I, 219–278.
59 WB, 6.
60 MT, 6–7.
61 ERE, 3–19.
62 ERE, 6–7.
63 ERE, 19.
64 James dedicated an entire lengthy notebook to this problem, addressing what he calls – after two of his critical interlocutors – the 'Miller-Bode objections' (MEN, 65–132).
65 LWJ II, 292.
66 PU, 154.
67 PU, 140.
68 PU, 94.
69 ERM 116–119; ERM, 105–108. Each of these works is a translation from an earlier German edition. Lutoslawski did not find a publisher for his English-language edition until 1924 – 14 years after James's death – at which point it was finally printed with James's 1899 introduction (Perry 1935 II, 214).
70 PU, 140–144.
71 PU, 145.
72 VRE, 412.

73 James was a founding member of the American Society for Psychical Research and was particularly impressed by a purported medium named Mrs Piper. See James's *Essays in Psychical Research*. For a popular account of psychical research among scientists during James's time – including natural selection's co-discoverer Alfred Russell Wallace – see Blum (2006). For James on abnormal psychology, see Taylor (1984).
74 VRE, 195.
75 ERM, 75–101.
76 ERM, 76.
77 PU, 142.

References

Barad, K. 2007. *Meeting the Universe Halfway: Quantum Physics and the Entanglement of Matter and Meaning*. Durham, NC: Duke University Press.
Bateson, G. 1972. *Steps to an Ecology of Mind*. Chicago, IL: University of Chicago Press.
Bergson, H. 1907. *L'évolution créatrice*. Paris: Les Presses universitaires de France.
Blood, B. P. 1893. *The Flaw in Supremacy*. Amsterdam, NY: Published by the author.
Blum, D. 2006. *Ghost Hunters: William James and the Search for Scientific Proof of Life after Death*. New York: Penguin Press.
Cummins, R. 1975. Functional Analysis. *Journal of Philosophy* 72, 741–764.
Dewey, J. 1896. The Reflex Arc Concept in Psychology. *Psychological Review* 3, no. 4: 357–370.
Fechner, G. T. 1904. *The Little Book of Life after Death*. Translated by M. C. Wadsworth. Boston, MA: Little, Brown, & Company. Original edition, 1836.
Gould, S. J. and R. C. Lewontin. 1979. The Spandrels of San Marco and the Panglossian Paradigm: A Critique of the Adaptationist Programme. *Proceedings of the Royal Society of London. Series B, Biological Sciences* 205, no. 1161: 581–598.
Gould, S. J. and E. S. Vrba. 1982. Exaptation-A Missing Term in the Science of Form. *Paleobiology* 8, no. 1: 4–15.
Haraway, D. J. 1991. Situated Knowledges: The Science Question in Feminism and the Privilege of Partial Perspective. In *Simians, Cyborgs, and Women: The Reinvention of Nature*, 183–202. New York: Routledge.
Harding, S. 1986. *The Science Question in Feminism*. Ithaca, NY: Cornell University Press.
James, H. 1884. *The Literary Remains of the Late Henry James*. Boston, MA: Houghton Mifflin.
James, W. 1975. *The Meaning of Truth*. The Works of William James. Edited by F. Burkhardt, F. Bowers and I. K. Skrupskelis. Cambridge, MA: Harvard University Press. Original edition, 1909.
James, W. 1975. *Pragmatism*. The Works of William James. Edited by F. Burkhardt, F. Bowers and I. K. Skrupskelis. Cambridge, MA: Harvard University Press. Original edition, 1907.
James, W. 1976. *Essays in Radical Empiricism*. The Works of William James. Edited by F. Burkhardt, F. Bowers and I. K. Skrupskelis. Cambridge, MA: Harvard University Press. Original edition, 1912.
James, W. 1977. *A Pluralistic Universe*. The Works of William James. Edited by F. Burkhardt, F. Bowers and I. K. Skrupskelis. Cambridge, MA: Harvard University Press. Original edition, 1908.
James, W. 1978. *Essays in Philosophy*. The Works of William James. Edited by F. Burkhardt, F. Bowers and I. K. Skrupskelis. Cambridge, MA: Harvard University Press.

James, W. 1979. *Some Problems of Philosophy*. The Works of William James. Edited by F. Burkhardt, F. Bowers and I. K. Skrupskelis. Cambridge, MA: Harvard University Press. Original edition, 1911.

James, W. 1979. *The Will to Believe and Other Essays in Popular Philosophy*. The Works of William James. Edited by F. Burkhardt, F. Bowers and I. K. Skrupskelis. Cambridge, MA: Harvard University Press. Original edition, 1897.

James, W. 1981. *The Principles of Psychology*. 2 vols. The Works of William James. Edited by F. Burkhardt, F. Bowers and I. K. Skrupskelis. Cambridge, MA: Harvard University Press. Original edition, 1890.

James, W. 1982. *Essays in Religion and Morality*. The Works of William James. Edited by F. Burkhardt, F. Bowers and I. K. Skrupskelis. Cambridge, MA: Harvard University Press.

James, W. 1983. *Essays in Psychology*. Edited by F. Burkhardt, F. Bowers and I. K. Skrupskelis. Cambridge, MA: Harvard University Press.

James, W. 1983. *Talks to Teachers on Psychology: And to Students on Some of Life's Ideals*. Edited by F. Burkhardt, F. Bowers and I. K. Skrupskelis. Cambridge, MA: Harvard University Press. Original edition, 1899.

James, W. 1985. *The Varieties of Religious Experience*. The Works of William James. Edited by F. Burkhardt, F. Bowers and I. K. Skrupskelis. Cambridge, MA: Harvard University Press. Original edition, 1902.

James, W. 1986. *Essays in Psychical Research*. The Works of William James. Edited by F. Burkhardt, F. Bowers and I. K. Skrupselis. Cambridge, MA: Harvard University Press.

James, W. 1988. *Manuscript Essays and Notes*. The Works of William James. Edited by F. Burkhardt, F. Bowers and I. K. Skrupskelis. Cambridge, MA: Harvard University Press.

James, W. 1988. *Manuscript Lectures*. The Works of William James. Edited by F. Burkhardt, F. Bowers and I. K. Skrupskelis. Cambridge, MA: Harvard University Press.

James, W. and H. James. 1920. *The Letters of William James: Two Volumes Combined*. Boston, MA: Little, Brown, and Co.

Lloyd, E. A. 1995. Objectivity and the Double Standard for Feminist Epistemologies. *Synthese* 104, no. 3: 351–381.

Longino, H. E. 1995. Gender, Politics, and the Theoretical Virtues. *Synthese* 104, no. 3: 383–397.

Lutoslawski, W. 1924. *The World of Souls*. London: George Allen and Unwin Ltd.

Mandelbaum, M. 1971. *History, Man, and Reason: A Study in Nineteenth-Century Thought*. Baltimore: Johns Hopkins Press.

Millikan, R. G. 1984. *Language, Thought, and Other Biological Categories*. Cambridge, MA: MIT Press.

Nietzsche, F.W. 1968. *On the Genealogy of Morals*. 2nd edn. In *The Basic Writings of Nietzsche*, 437–599. Translated by W. Kaufmann. New York: Modern Library. Original edition, 1887.

Nietzsche, F.W. 1974. *The Gay Science*. 2nd edn. Translated by W. Kaufmann. New York: Vintage Books. Original edition, 1882.

Oyama, S. 2000. *The Ontogeny of Information: Developmental Systems and Evolution*. 2nd edn. Science and Cultural Theory. Edited by B. H. Smith and R. E. Weintraub. Durham, NC: Duke University Press. Original edition, 1985.

Peirce, C. S. 1878. How to Make Our Ideas Clear. *Popular Science Monthly* 12, 286–302.

Perry, R. B. 1935. *The Thought and Character of William James*. 2 vols. Boston, MA: Little, Brown, and Company.

Seigfried, C. H. 1984. Extending the Darwinian Model: James's Struggle with Royce and Spencer. *Idealistic Studies* 14, no. 3: 259–272.

Seigfried, C. H. 1991. Where Are All the Pragmatist Feminists? *Hypatia* 6, no. 2: 1–20.

Seigfried, C. H. 1996. *Pragmatism and Feminism: Reweaving the Social Fabric*. Chicago, IL: University of Chicago Press.

Sober, E. 1984. *The Nature of Selection*. Cambridge, MA: MIT Press.

Suckiel, E. K. 1982. *The Pragmatic Philosophy of William James*. Notre Dame, IN: University of Notre Dame Press.

Sullivan, S. 2001. *Living Across and Through Skins: Transactional Bodies, Pragmatism and Feminism*. Bloomington, IN: Indiana University Press.

Taylor, E. 1984. *William James on Exceptional Mental States: The 1896 Lowell lectures*. New York: Scribner.

Thayer, H. S. 1975. Introduction. In *The Meaning of Truth*, eds F. Burkhardt, F. Bowers and I. K. Skrupselis, xi–xlvi. Cambridge, MA: Harvard University Press.

Williams, B. 2004. *Truth and Truthfulness: An Essay in Genealogy*. Princeton, NJ: Princeton University Press.

Winther, R. G. In preparation. *When Maps Become the World*. Chicago, IL: University of Chicago Press.

Wright, L. 1973. Functions. *Philosophical Review* 82, 139–168.

Conclusion
Divided selves and dialectical selves

Natural selection is not only a revolutionary scientific idea. It is also a philosophical bombshell on par with Nietzsche's proclamation that God is dead. Evolutionism in general means that nature is changing rather than static, and indeed that stability requires explanation. Darwinism in particular challenges further ancient and deeply held beliefs: that humans enjoy a privileged status within an ontological hierarchy of nature; that large-scale ordered change is the result of benevolent foresight; that the organic world expresses ideal essences; and that science advances progressively through simple induction from observation. Human purpose and knowledge are historicized, decentred and ontologically deflated. We are not quite at home in the world.

The Received Image of Darwinism makes sense of this situation by assimilating evolutionism to an Enlightenment philosophy of mechanistic determinism. In doing so, it supports a philosophy of 'nothing but': Organisms are nothing but a resultant of genetic programmes triggered by external factors; and populations are nothing but a product of the mechanical sorting of mutations by an environment. The proximate story is preformationist, while the ultimate story is externalist. Neither story features the developing organism as such. As Richard Lewontin has remarked, this elision of the individual in neo-Darwinian biology is ironic, given that Darwin was inspired by classical economic theories that enshrine the individual as the chief locus of value, responsibility and causal efficacy.[1] If the anglophone economic individualism that inspired Darwin reached its apotheosis in the rhetoric and policies of the Reagan and Thatcher administrations of the 1980s, then by this same period neo-Darwinism had practically dissolved the individual at the intersection of genes and environment.

The point of this book has not been to support economic individualism or social Darwinism, but it has aimed to reconstitute individuality in a post-Darwinian worldview through an examination of William James's Pragmatic Image of Darwinism. The latter worldview offers a thoroughgoing reconstruction of purpose, knowledge and reality, characterized by a commitment to internalism and constructionism in organic systems; generalized selectionism within the sensorimotor system and beyond; fallible knowledge and indeterminate truth; and a conception of reality as an open-ended, dynamically continuous system. With the partial exception of generalized selectionism, these positions are not definitive of neo-Darwinism. Nevertheless, they grew organically out of James's work in Darwinian physiology and psychology. They are thus *Darwinian* ways of responding to Darwinism.

James trained as a physiologist and medical doctor at Harvard in the 1860s, in the immediate wake of Darwin's *Origin of Species*. James's earliest publications show him moving toward Jeffries Wyman and Asa Gray's provisional acceptance of natural selection as a modern kind of fallible science and away from Louis Agassiz's idealistic conception of organic form. Harvard also provided the setting of the Metaphysical Club in the 1870s, where Charles S. Peirce and Chauncey Wright broke James of his youthful enthusiasm for the teleological Lamarckism of Herbert Spencer. These intellectual exchanges were supplemented by James's own readings. Charles Renouvier in particular inspired James's philosophy of pluralism and his concept of will as selective attention. By the middle of the 1870s, James's project can be understood as a synthesis that he himself prescribes to the scientific community in an early book review:[2] that of Renouvier's doctrine of will and Bain and Carpenter's conception of moral practice as the cultivation of plastic habits.

Within this constellation of influences, Darwin's theory of natural selection remains the logical backbone of James's psychology and philosophy. This is because Darwin gives James the idea of non-directed variation, or variation that is not produced or elicited by external conditions. James regularly invokes non-directed variation in challenging externalist accounts of ontogeny and phylogeny, as well as in construing higher-order systems on analogy with ontogeny. This logic even informs James's adaptation of Renouvier's conception of will, since the will for James is unable to produce the ideas that it selects. Indeed, moral action is defined by James as the mediation of extant structures by novelty that is fed to the will from elsewhere. The point here is not that selection explains everything, or that selectionist processes are the only significant factors in complex systems. It is that James finds selectionism useful for capturing the nonlinear nature of numerous systems in phylogeny, ontogeny, society and history. Selectionism for James is emblematic of his pluralistic thesis that different parts of nature may be connected only by way of intermediary connections, rather than holding together in a 'block universe' or 'total conflux'.

James built an ethics of self-transformation upon this Darwinian structure. If selectionism denies the environment the power to produce variation, it nevertheless allows the environment to alter the structures that produce variation in the future. The environment thus exercises a kind of indirect influence over variation, such that today's selections bias the conditions of tomorrow's variation. Unlike natural selection, however, the Jamesian will is able to exert this influence in an intelligent and purposive manner. This is what occurs when one engages in self-transformation, or promotes certain of one's interests at the expense of others: The habitual environment is altered, changing the conditions for future variation. Indeed, the crux of James's ethics and his entire melioristic philosophy is that individuals may in this way spiral their ideals both centripetally into themselves and centrifugally into a broader cooperative social world.

James's moral theory is thus not an act-based one like Kantianism or utilitarianism, but a character-based one that aims at flourishing or eudemonia. It could thus be described as Aristotelian or perhaps existentialist.[3] Better, however, is to avoid the impulse to pigeonhole and to recognize James as an idiosyncratic thinker whose originality lay largely in his creative exaptation of concepts from physiology and evolutionary theory. Nietzsche's philosophy can be described in much the same way, even if he devalues social cohesion and prizes ascetic self-overcoming more than a hopeful 'willing-to-believe'. If the danger of James's view is wishful thinking, the danger of Nietzsche's view is alienation. The choice between these visions is a pragmatic one with no pre-given answer.

The framework of self-transformation places James's views on belief, truth and reality in a different light. 'The Will to Believe', for instance, is not an isolated contribution to a niche ethical debate but a case study in self-transformation in which James posits the right to develop oneself according to the energies opened up through acts of will. When James expands Peirce's pragmatic maxim to include effects on conduct, moreover, he is interpreting both meaning and truth in terms of the sensorimotor dialectic that constitutes the self-mediating individual as such. Finally, James proposes that the world itself is an evolving system consisting of a cascading hierarchy of increasingly inclusive selves: We are multi-layered developing organisms whose awareness is fringed by an inchoate 'more' that is never quite grasped; and we reside in a similarly multi-layered, fringed and growing world, through which is woven a living tissue of truth.

Prospects for a pragmatic Darwinism

James's Pragmatic Image of Darwinism has after-images in the twentieth and twenty-first centuries. It is worth taking stock of these images to assess the prospects for a pragmatic Darwinism today. This concluding discussion highlights relevant schools of thought in relation to one of the key concepts of James's later philosophy: *vicious abstractionism*, or the method of treating a concept or theory as exhaustive of the reality that it purports to explain. One antidote to vicious abstractionism, emphasized below, is *dialectical thinking*. James was ambivalent toward dialectics in its monistic guise, but a dialectical logic captures something of the dynamic and nonlinear quality that he was attempting to get at in his selectionism.

Vicious abstractionism

James is not against concepts, which allow us to 'hop, skip and jump over the surface of life at a vastly rapider rate than if we merely waded through the thickness of particulars as accident rained them down upon our heads'.[4] He is on guard, however, against those who would make the flux of experience ontologically derivative of the conceptual. James vacillates in exactly how deflationary he wants to be about concepts. When he is at his most Bergsonian (as in *A Pluralistic Universe*), he claims that concepts inherently misrepresent the flux of reality. He is more even-handed, however, in his final unfinished work *Some Problems of Philosophy*:

> Concepts are thus as real as percepts, for we cannot live a moment without taking account of them. But the 'eternal' kind of being they enjoy is inferior to the temporal kind, because it is so static and schematic and lacks so many characters that temporal reality possesses.... Perceptual reality involves and contains all these ideal systems, and many more besides.[5]

Concepts are real for the pragmatic reason that we must constantly negotiate a world schematized according to their logical relations. Nevertheless, this 'eternal' world is revisable and derives from the variegated temporal world that embeds it.

In this context, vicious abstractionism means using concepts 'privatively' for the sake of a false simplicity:

> We conceive a concrete situation by singling out some salient or important feature in it, and classing it under that; then instead of adding to its previous characters all the positive consequences which the new way of conceiving it may bring, we proceed to use our concept privatively; reducing the originally rich phenomenon to the naked suggestions of that name abstractly taken, treating it as a case of 'nothing but' that concept, and acting as if all the other characters from out of which the concept is abstracted were expunged.[6]

Vicious abstractionism is a sort of epistemic bait-and-switch in which one posits a skeletal abstraction as uniquely explanatory of the reality from which it was abstracted. According to James, this abuse of concepts is 'one of the great original sins of the rationalistic mind'.[7] James traces this problem back to the ancient Greeks:

> Ever since Socrates we have been taught that reality consists of essences, not of appearances, and that the essences of things are known whenever we know their definitions. So first we identify the thing with a concept and then we identify the concept with a definition, and only then, inasmuch as the thing *is* whatever the definition expresses, are we sure of apprehending the real essence of it or the full truth about it.[8]

In contrast, James's pluralism holds that a particular categorization does not exclude the possibility of other categorizations, since no category is exhaustive of the reality it describes.

Vicious abstractionism belongs to a family of fallacies described by James and Dewey, which together comprise a robust pragmatist critique of abstraction. Rasmus Grønfeldt Winther has brought these fallacies together under the banner of 'pernicious reification'.[9] If abstraction means singling out aspects of the world, symbolizing them and systematizing these symbols, then different kinds of pernicious reification result from the ignoring of different kinds of context in different phases of this process. For instance, one may ignore functional context, historical context or context regarding the appropriate analytic level at which explanations are pitched. The elision of context is pernicious because it results in the dubious universalizing, narrowing or ontologizing of theories: One's pet theory explains too much; or it renders meaningless phenomena that it cannot capture; or it is ontologically inflated so as to serve as a stand-in for the world that it represents. These problems are arguably exhibited, for instance, by Richard Dawkins's theory of the selfish gene.[10] Winther's prescription for pernicious reification includes theoretical pluralism and the critique of assumptions in their functional and historical context ('assumption archaeology'). Armed with such tools, theorists may view disparate models as complementary rather than competing, both across and within fields.[11]

The discussion here will focus in particular on combatting vicious abstractionism through dialectical thinking.

Hegel's dialectic

In James's opinion, the worst perpetrator of vicious abstractionism is not Socrates or Plato but Hegel. James specifically critiques Hegel's logic,[12] which is a radical development of the transcendental logic of Kant's *Critique of Pure Reason*. Unlike Kant's formal logic, which concerns the form of thought only, transcendental logic concerns the way in which the empirical world comes to be given. Hegel broadens the role of transcendental logic from that of producing objects for finite knowers to producing knowledge absolutely. Thus, whereas transcendental logic for Kant is about how finite intellects constitute objects of knowledge in the delimited phenomenal realm only, logic for Hegel is about how the world itself develops and thereby expresses the 'Absolute Idea'.[13] Logic for Hegel is therefore the science of the self-development of thinking, which is also the development of the world as it generates increasingly adequate expressions of the Idea. Since the form and content of thought are inseparable for Hegel, he claims that his logic captures 'the absolute form of the truth and, even more than that, also the pure truth itself'.[14]

A particular structure of three logical moments or 'sides' is ubiquitous in Hegel's system. This structure is represented by what Hegel calls the 'speculative method' (often later called the 'dialectic method'):[15]

1   The side of abstraction or the understanding.
2   The dialectical or negatively rational side.
3   The speculative or positively rational side.

In the first moment, the understanding attempts to grasp a specific determination among others in order to get at its truth immediately. This effort is frustrated because the full truth of any determination is shown only to exist through other, opposing determinations. This necessitates the second moment, which is 'the self-sublation of these finite determinations on their own part, and their passing into their opposites'.[16] This is the dialectical moment in which anything finite is shown to imply its own lack of self-sufficiency and its need to pass into its own negation in order to express its truth. Finally, in the speculative moment, the opposing determinations are brought together under a higher unity, which is 'positively rational' in that it realizes the fuller truth implicit in the apparent opposition.

These three logical moments structure Hegel's system at each level, ranging from sensation to the overarching journey of the Absolute. This highest-level journey can be seen in the names of the three volumes of Hegel's *Encyclopaedia of Philosophical Sciences*: Through its self-developing thinking (*Logic*), the Idea comes to other itself in nature (*Nature*), before coming back home to itself fully realized (*Spirit*). This ultimate application of the triadic logic is peculiar in that the Absolute has no external other in relation to which it can be negated. The second moment of the Absolute's development therefore requires an act of *self*-othering, which sounds like a difficult metaphysical contortion. In the end, the truth of any given determination is fully expressed only in the fully realized Absolute Idea. That everything real is only real in this manner is the meaning of Hegel's objective idealism, summed up in the claim that everything is 'for' the Absolute.

James credits Hegel with introducing to anglophone philosophy a certain 'expansion and freedom' from classical empiricism,[17] due to his vision of a 'dialectic movement in things' where 'whatever equilibriums our finite experiences attain to are but provisional'.[18] He thus argues that Hegel's logic captures a genuine insight:

> Take any concrete finite thing and try to hold it fast. You cannot, for so held, it proves not to be concrete at all, but an arbitrary extract or abstract which you have made from the remainder of empirical reality. The rest of things invades and overflows both it and you together, and defeats your rash attempt. Any partial view whatever of the world tears the part out of its relations, leaves out some truth concerning it, is untrue of it, falsifies it.... Taken so far, and taken in the rough, Hegel is not only harmless, but accurate. There *is* a dialectic movement in things, if such it please you to call it.[19]

Manifestly, there is dialectical mediation of concepts in experience. New concepts may always enlarge one's perspective or demonstrate the limits of prior concepts. Indeed, concepts 'invade' even when one tries to understand a particular or property in isolation. However, James accuses Hegel of vicious abstractionism (or 'vicious intellectualism') because of his characterization of this process:

> Now Hegel himself, in building up his method of double-negation, offers the vividest possible example of this vicious intellectualism. Every idea of a

finite thing is of course a concept of *that* thing and not a concept of anything else. But Hegel treats this not being a concept of anything else as if it were *equivalent to the concept of anything else not being*, or in other words as if it were a denial or negation of everything else. Then, as the other things, thus implicitly contradicted by the thing first conceived, also by the same law contradict *it*, the pulse of the dialectic commences to beat and the famous triads begin to grind out the cosmos.[20]

Strikingly, it would seem that Hegel's logic *just is* vicious abstractionism, posited as the very engine of the dialectic. Thought for Hegel is always over-reaching itself, as concepts pretend that they may self-sufficiently capture a kind of truth that necessarily exceeds them. Hegel needs concepts to puff themselves up in this way to get his metaphysical machinery running.

James's critique is unfair, however, because it conflates levels of analysis. It is true that Hegel posits a logic in which the understanding commits vicious abstractionism by attempting to single out a determination and know it immediately (assuming that a fallacy can be attributed to a *logical moment*). However, Hegel's entire point is that the understanding is inferior to the positively rational moment in which the determination has been mediated by its opposite. In fact, Hegel argues in 'Sense-Certainty' that immediate sensation simply does not exist. Immediacy in the sense demanded by the understanding is a ruse, as can be demonstrated by any attempt to grasp a mere 'This' without concepts.[21] This means that Hegel is not guilty of positing discrete sense data and then bringing in the Absolute as a *deus ex machina* to stitch them together – as James sometimes suggests. On the contrary, Hegel is critiquing the logical moment of the understanding in much the same way that James is critiquing the viciously abstractionist philosopher. The vicious abstractionism of Hegel's logical moments should not be attributed to Hegel himself.

If Hegel commits vicious abstractionism, it is not because of his dialectical method per se but because he uses this method to drive a totalizing rational system that is supposed to encompass a uniquely correct account of all truth and reality. Hegel tells us that the truth (without remainder) is the whole, and he outlines the whole for us. Thus, for all of the historicism of the *Phenomenology of Spirit*, Hegel is open to the same fundamental objection as all the idealists who shattered the critical limits set by Kant: He claims a vantage point beyond finite thought and from there makes totalizing claims that cannot be engaged from outside of his system. (Hegel would, of course, deny that he has transcended the Absolute, claiming instead that the Absolute is realizing itself *through him*.) After a few decades of such philosophy, one can imagine the sardonic enjoyment of a James or a Nietzsche casting such self-important systems as functions of biography or temperament.

James's actual difference from Hegel comes out in his report on pondering the latter's system after inhaling nitrous oxide (an experiment that he recommends).[22] According to James, the gas allowed him to experience a great 'metaphysical illumination' in which concepts seemed to negate one another and interpenetrate. Upon recovering, however, James concluded that Hegel's dialectic 'is really a self-consuming process, passing from less to the more abstract, and terminating either in a laugh at the ultimate nothingness, or in a mood of vertiginous amazement at a meaningless infinity'.[23] This is what James rejects: an upward-spiralling dialectic of increasing abstraction. Hegel's Absolute Idea is the apotheosis of contentless abstraction.

It is a shame that James did not see more potential in dialectical thinking, which he associated strongly with his monistic interlocutors. Other forms of dialectic remain closer to the ground.

Dialectical biology and developmental systems theory

If Hegel's idealism rejects the reductionistic mechanism of the Enlightenment, then dialectical biology is a rejection of the twentieth-century incarnation of such a view in neo-Darwinism. Dialectical biology is the view championed by Richard Levins and Richard Lewontin in *The Dialectical Biologist* (1985). Dialectical biology lies closer to the Marxian or materialistic wing of dialectical philosophy than it does to Hegel – keeping in mind that this is a 'materialism' on which matter is inherently dynamic rather than inert. Instead of following the Cartesian method of isolating discrete atomistic units in order to model their interactions, dialectical biology assumes that units can only be understood in terms of the relations in which they participate. It therefore does not privilege a reductive 'basement' level of analysis in either biology or physics, arguing instead for a hierarchical view on which the organism as such may count as one locus of agency among others.

A central claim of Levins and Lewontin's book is that the organism is both a subject and object of evolution. This means that an environment (or niche) is not an autonomous coercive power but is itself constructed (in part) by the activities of its organisms. In other words, evolution is not the increasingly accurate moulding of a key (organism) to fit a lock (environment).[24] On the contrary, organism and environment redirect each other in an endless dynamic feedback process where stability is a dynamic achievement. Given this framework, Lewontin suggests that the metaphor of *adaptation* should be replaced by that of *construction*, in a dual sense: Organisms construct their environments concretely in physically altering them (for instance, by consuming resources and building homes); but they also construct their environments *conceptually* in that the environment may only be specified with reference to what is salient to the organism. The world does not contain abstract niches waiting to be filled.[25]

Dialectical biology is closely allied with another biological programme that is critical of standard neo-Darwinian assumptions: developmental systems theory. The birth of developmental systems theory can be dated to Susan Oyama's *The Ontogeny of Information* (1985), which advances a powerful critique of the modern concept of genetic information. Although it is admitted that the external world or developing body can modulate the genetic code, biologists grant genetic information a special causal status. Other factors are viewed as interference, or at best as background conditions. Oyama understands this conception of information as a modern manifestation of a long metaphysical tradition that construes matters as inherently inert and thus in need of dynamizing by an active and intentional force. If a God or Prime Mover provides this function for the cosmos as a whole, then genetic information serves a similar function at the proximate level by conferring form upon organisms. Therefore, genes are not just more heavily weighted than other causes of ontogeny; they are metaphysically exalted in a heady conflation of material DNA and ethereal information.

As both Lewontin and Oyama argue, this concept of information is today's version of *preformationism* – the old view that a tiny individual exists inside either the sperm or the egg such that ontogeny consists in *getting larger*. Although this sounds silly to modern ears, the modern concept of genetic information implies with the preformationists that the mature individual is somehow 'already in there'. Against this conception of information, Oyama suggests a full conceptual about-face: Information does not pre-exist its expression; information is just a 'difference that makes a difference' in the contingent and multi-causal processes of development. In other words, information has an ontogeny. For biology to take this critique on board would constitute a 'stake-in-the-heart move', where 'the heart is the notion that some influences are more equal than others, that form – or its modern agent, information – exists before the interactions in which it appears'.[26]

This democratizing of the causes of ontogeny encourages an extended view of heredity or inheritance. Heredity need not refer – after neo-Darwinism (or neo-Weismannism) – only to that which is transmitted in the genes. In the broadest sense, heredity refers to the simple fact that offspring have more in common with their parents than with the general population. The fact of heredity and its explanation are two different things. According to developmental systems theory, genes are one developmental resource among others that are repeatedly made available across life cycles. This site of developmental resources includes 'chromosomes, nutrients, ambient temperatures, childcare … chromatin marks that regulate gene expression, cytoplasmic chemical gradients and gut- and other endosymbionts', in addition to 'the local physical environment, altered by past generations of the same species and other species as well as the organism's own activities'.[27] It is misleading to claim that genes hold our potential, where our environments determine how that potential is unlocked. It is equally true that the myriad inherited structures of the wider social-natural world hold our potential and that is up to our *genes* not to mess this up. Both explanations are one-sided, even if the former way of talking is now entrenched to the point of being common sense.

Dialectical biology and developmental systems theory are consonant with James's Pragmatic Image of Darwinism. In his psychology, James rails against the idea that organisms adapt passively to environments (an idea he associates more with Lamarckism than Darwinism). Moreover, James's radical empiricism fundamentally agrees with Oyama that matter (as one aspect of pure experience) is inherently dynamic and therefore does not require any external agent to torque it into motion. In addition, while he emphasizes the internal causes of ontogeny (as a way of combatting externalism), James views such causes in a radically non-preformationist fashion. Variation for James is inchoate fodder, which has no meaning or function until it is mediated by the developing organism. James even defines truth itself in terms of the unfolding of differences that make a difference in experience, such that 'absolute truth' itself has no predetermined shape. Experience for James is finitely rational in its dynamic continuity, where nothing is there before it is there.

Autopoesis: dual dialectics

A Jamesian dialectical biology would encompass a dialectical psychology, where mind is understood in terms of life. The point here is not to reduce psychology to biology but to follow Spencer's lead in viewing mind as a manifestation of the logics and processes that define life in general.

Selectionism is an important part of this story for James: A double-barrelled Darwinian psychologist, James posits selectionist processes in both the phylogeny and ontogeny of mind. Selection by itself may be taken to pernicious extremes, however. The extreme behaviourist, for instance, construes individuals as interchangeable blank slates inscribed by trial-and-error conditioning. This is far from James's intention. Selectionism may even provide a way of discharging purpose from the world entirely. This is why neo-Darwinian philosopher Daniel Dennett, for instance, gives pride of place to the Law of Effect – Edward Thorndike's formalization of trial and error – in psychology. Any account of intelligence or purpose cannot itself invoke intelligence or purpose, on pain of circularity. Dennett thus argues that the Law of Effect, as a mechanistic account of apparent purpose, must be assumed a priori by any psychological theory worthy of the name.[28] Just as natural selection discharges purpose from phylogeny, the Law of Effect discharges purpose from ontogeny. Purpose everywhere must bottom out in mechanistic trial and error.

James agrees with Dennett that interests spring ultimately from non-intentional physiological processes. He therefore does not suffer from the infinite regress of homunculi that is Dennett's concern. Nevertheless, he contends that the individual as such acts purposively. If purposiveness is a system-level property of the organism, then it need not be chased to ever lower levels of analysis until it is finally expelled as a fiction. It is a strong – and viciously abstractionist – presumption that eliminating the higher through the lower provides a uniquely accurate account of living beings. Purpose may disappear when one squints from a certain standpoint. However, the same is true of any given property of a system. This provides a major *disanalogy* between natural selection and intelligent behaviour for James: The latter is genuinely purposive while the former is not. Evolution accidentally made beings that do things on purpose.

A more refined analysis of life and mind goes beyond James's talk of variation and selection – so beloved of externalists and reductionists – and employs a more dialectical vocabulary. If life and mind for Spencer are characterized by hapless adaptation to a coercive environment, and if life and mind for Dennett are characterized as a pseudo-purposiveness generated by mechanistic selection, then for dialectical theorists Humberto Maturana and Francisco Varela the essence of life and mind is *autopoesis*.[29] Autopoesis means the self-organization of a dynamic pattern that is maintained even as the specific materials (such as cells or tissues) may be destroyed or replaced. According to Varela,

> [a]n autopoetic system is organized (defined as unity) as a network of pro-
> cesses of production (synthesis and destruction) of components such that
> these components:
>
> i continuously regenerate and realize the network that produces
>   them, and
> ii constitute the system as a distinguishable unity in the domain in which
>   they exist.[30]

This particular form of self-organization is constituted at the intersection of two dialectics: first, a part–whole dialectic in which the organism is constituted as a self-enclosed system; and second, a self-other dialectic in which the organism is coordinated with a lived world or *Umwelt*. These two processes capture the basic dialectical idea that something is what it is through its opposition – conceptual and/or physical – to some other. Parts are only parts as parts-of-a-whole, just as a whole is only a whole as a whole-of-parts. (A refusal to logically privilege the Whole is what makes this dialectical non-Hegelian.) Similarly, although the organism's *Umwelt* is defined as an external other, it is specifically the external other *of the organism* as opposed to an arbitrary or unrelated other. To be a self is to be already invaded by an encroaching non-self. This situation is not to be decried, as there is no other kind of identity to be had: Identity is a provisional accomplishment held in place through multiple interwoven tensions.

In this context, the self may be understood as a virtual locus within a shifting cluster of 'regional selves' in different domains. For human beings, this includes an immunological self,[31] a sensorimotor self, the subjective 'I' invoked in language and others.[32] According to Varela, it is useful to view these different selves in relation to one another rather than to treat them as discrete. Indeed, James had already arranged multiple 'regional' selves in a hierarchy in his chapter 'The Consciousness of Self' in *The Principles of Psychology*. The present study has focussed on the sensorimotor self in James, that is, the self as it is constituted by the reflex arc (or *circuit*) of sensation-cognition-action. The closure of the sensorimotor circuit confers cognitive and perceptual coherence on experience, while constituting the external *Umwelt* as both a site of knowledge and a stage for dramatic action. Other layers of selfhood could be further investigated in relation to this sensorimotor conception. For instance, the social dimension of identity has scarcely been touched upon in the present study – except insofar as society itself has been modelled on analogy with a developing sensorimotor self.

Autopoesis makes life purposive by definition. Once there is a simple bacterial cell, there is purpose. This does not mean that the cell is expressing an ideal species essence, or that it is advancing history according to a cosmic teleology. Nor does it mean that the cell is successfully doing what its ancestors were selected for doing in the evolutionary past (as in theories of *teleonomy*). The point is simply that the maintenance of an autopoetic system in relation to an *Umwelt*, in and of itself, constitutes a finite natural teleology. Organism and purpose are in this sense coeval, even if the form of this purposiveness may modulate with different forms of life (and their different imbricated regional selves). Whether a system higher than the physiologically integrated organism could be said to exhibit such natural purposiveness – for instance, a social group, or James's postulated hierarchy of minds – would depend on whether it satisfies the criteria for autopoesis. Differences between domains and levels of analysis must be respected. For the familiar organism, at least, the autopoetic view contradicts both Dennett's reductionism and the position attributed to Kant in the *Critique of Judgment* that purposiveness is a non-explanatory heuristic that we cannot help but impose on organisms.[33]

Humanistic psychology

A final ally against vicious abstractionism is the humanistic psychology movement of the middle of the twentieth century. This programme positioned itself as a 'Third Psychology', that is, an alternative to mechanistic behaviourism and Freudian psychodynamic theory that includes elements of both. Humanistic psychology studied the holistic conditions for human flourishing, rather than offering a negative conception of mental health as the mere removal of symptoms. It was also overtly pragmatic in that it aimed to produce concrete results through its influence on 'helping professions' such as psychotherapy, social work and education.

There are actual lines of influence between James's thinking and humanistic psychology, which have only rarely been explored.[34] James's most influential work in psychology – *The Principles of Psychology* (1890) – is closer to cognitive and behavioural science than it is to humanistic psychology. James also worked in abnormal psychology and psychodynamic theory, however, as evidenced in his 1896 Lowell Lectures 'On Exceptional Mental States' and in the psychotherapeutic dimensions of *The Varieties of Religious Experience* (1902).[35] As John McDermott has argued, moreover, James's metaphysics of radical empiricism portrays the inchoate fringes of experience as raw material for growth.[36] According to James in 'A World of Pure Experience', for instance,

experience itself, taken at large, can grow by its edges.... Life is in the transitions as much as in the terms connected; often, indeed, it seems to be there more emphatically, as if our spurts and sallies forward were the real firing-line of the battle, were like the thin line of flame advancing across the dry autumnal field which the farmer proceeds to burn.[37]

James thus took personality and growth as central organizing principles of his work. This orientation influenced the Harvard 'personality psychologists' Gordon Allport and Henry Murray in the early twentieth century,[38] who in turn – along with strains of existential philosophy and Gestalt psychology – influenced mid-century humanistic psychology. By the century's end, however, the whole individual in academic psychology had gone the way of the whole individual in biology, as humanistic psychology faded and was succeeded by its counter-cultural offspring transpersonal psychology.[39] Today, the strong focus on cognitive neuroscience in academic psychology does not encourage a person-centred science.[40]

Questions of influence aside, humanistic psychology is Jamesian in both its methodology and its concerns. This is particularly true of the work of Abraham Maslow, who will be the focus here. Maslow develops an idea from Carl Rogers and Kurt Goldstein, which is also nascent in James: Individuals exhibit a tendency, not just toward the maintenance of a homeostatic equilibrium, but also toward growth. Indeed, such growth may require the repeated deconstruction and reconstruction of a relatively stable self so as to achieve an increasingly satisfactory hierarchy of values.[41] James and Maslow may talk about hierarchies in different ways: If James posits a hierarchy among sub-personal 'centres of energy' representing possible modes of being (alternative selves), then Maslow famously posits a *hierarchy of needs* in which the satisfaction of lower needs is requisite for the satisfaction of higher ones. These two hierarchies are related, however, in that Jamesian self-transformation will be hampered to the extent that one's lower Maslovian needs are unfulfilled. After all, self-actualization lies at the very top of the Maslovian pyramid, above safety, belongingness, love and esteem.[42] Self-transformation ideally comes from a place of relative health that is precluded by deep-seated insecurity or alienation. This is not good news for the prototypical alienated existentialist forging a self in the face of a hostile world. There is something adolescent and masculinist in the desire to make oneself from nothing, as if one had no body, history or environment.

Maslow depicts the basic dialectic of human experience as that between safety and growth, or fear and courage.[43] The dialectical nature of this relationship can be seen in Maslow's description of neurosis as 'a clumsy groping forward toward health and toward fullest humanness, in a kind of timid and weak way, under the aegis of fear rather than courage'.[44] Even pathological behaviour is a compromised version of health, which may represent an understandable response to pathological circumstances (if, for example, it is legitimately unrealistic to feel safe or loved). Maslow's distinction between the aegis of fear and the aegis of courage mirrors the distinction in James's *Talks to Teachers on Psychology* (borrowed from Spinoza) between acting *sub specie mali* (under the aspect of the bad) and *sub specie boni* (under the aspect of the good).[45] If one is not bent into a negative posture by the frustration of basic needs, then one moves positively in the direction of growth. Humanistic psychology is thus an optimistic theory of human nature: Human beings naturally unfold toward self-actualization, even if they may be deflected on the way.

Marks of self-actualization for Maslow include the ability to be fully absorbed in a task; a tendency toward growth-oriented decisions rather than fear-oriented ones; expressing spontaneous impulses instead of indulging in phoniness (as in giving honest and unlearned reactions to, say, a glass of wine or a piece of literature); and abandoning short-sighted defences.[46] A high degree of self-actualization is uncommon, although each of us experiences it fleetingly in 'peak experiences'.[47] A striking conclusion here is that *health is not normal*. To identify health with the average is to mistake widely shared neuroses – for instance, the lack-driven need for incrementally higher degrees of material wealth or professional status – with an optimal state of being. According to humanistic psychology, the point in education and social planning should not be to produce individuals who are 'well-adjusted' to the average but to lift the average itself. It is in this sense an inescapably normative school of thought.

Humanistic psychology is susceptible to the critique that it is universalist and preformationist about human nature. To paraphrase Leo Tolstoy, Maslow seems to claim that – while different individuals may be unhappy in different ways – there is just one way to be happy. By Maslow's own description, self-actualization is supra-historical and unfolds from within.[48] On a charitable reading, Maslow's seeming universalism and preformationism may be read as a defence against asymmetric externalism rather than an attempt to enthrone an equally one-sided internalism on which the self-actualized person is *already in there somewhere*. One must recognize, as James does, that both flourishing and moral value are relative to an environment. Insofar as self-actualization is the same in everyone, this must signify a shared similarity in bio-psychological, geographical and cultural conditions among all *Homo sapiens*. The extent to which such conditions are shared is a vexed question with no agreed-upon answer. Perhaps there are common patterns of self-actualization that are consistently reconstructed in ontogeny, which constitute a finite teleology defined in dialectical opposition to both internal neuroses and their allies in pathological familial and social systems. Descriptively speaking, such a self-actualizing tendency is universal across individuals and cultures just to the extent that it is. Normatively speaking, to posit the fostering of such self-actualization as the very goal of society is Maslow's utopian vision – just as the community of 'saints' in *The Varieties of Religious Experience* is James's utopian vision.[49]

Humanistic psychology is of particular interest because it addresses the role of abstraction in both science and self-actualization. Maslow treats these topics in *The Psychology of Science* (1966), an examination of science as a product of human beings.[50] As in James's pragmatism, Maslow intends to include and enlarge upon the extant sciences so as to provide a broader framework from which to assess and critique the production of knowledge. Such a critique foregrounds the human implications of all science, and especially sciences like psychology that take human beings as their subject of study. This means critiquing scientific practice from the perspective of the whole hierarchy of needs. According to Maslow, science serves cognitive needs and needs for safety and control, although it may also express deficiencies in these needs.[51]

In this holistic context, abstraction is a double-edged sword: Scientific abstraction rubricizes experience so that it may be handled according to certain purposes; but this very process simplifies and may foreclose upon alternative understandings. The latter problem is exacerbated with relatively successful research programmes like neo-Darwinism or (in a different time) Newtonian physics. Especially for those whose careers depend upon the continued implementation of particular methodologies, there is a tendency to cast as meaningless any phenomena that these tools do not handle well. Indeed, Maslow's *Psychology of Science* helped give currency to the adage, '[i]f all you have is a hammer, everything looks like a nail'.[52] In contrast, Maslow recommends orienting oneself to a *problem* that may be approached through any number of different methodologies.[53]

Even within a given methodology, however, one must be wary of the tendency to reify distinctions into dichotomies. It is a consistent theme of Maslow's writings that dichotomizing is both a cause and effect of pathology; or, to put it positively, it is a mark of health to understand opposites as dialectically entwined poles that may be superseded from a higher perspective.[54] For instance, the idea that evolution proceeds by way of an interaction between genes and environment is elegantly simple, but for this very reason it loses traction with a world in which genes are embedded in developing and niche-constructing organisms. Gene and environment are literally unthinkable without the developmental systems that define them as such. The point is not that evolution may not be modelled in simplifying ways. Scientific theory cannot be a 1:1 map of the world.[55] The important thing is one's attitude toward abstraction. Healthy or generative abstraction is a tool for re-engaging the potentially radical and surprising qualities of the concrete. An essential function of general knowledge is to illuminate any given particular. If the idiographic is steamrolled by the nomothetic, then something has gone wrong.

These issues take on a particular ethical cast in the classification of human beings, which Ian Hacking calls 'interactive kinds' because they alter their behaviour in response to being labelled.[56] In James's idiom, a viciously abstractionist attitude toward oneself compromises one's self-transformation. In his notes from 1899 to 1901, for instance, James depicts a kind of compromised self-management *sub specie mali*:

> So a man to 'save' himself, can throw himself in turn on this or that one of the functions or aspects of character in which he has least failed, and treat that as if it were the essence for which alone he should be judged and held responsible. He can choose which of his *mes* to take a stand on, and not count others; enter into personal relations which entirely abstract from others the existence, loathing and pitying them from the higher point of view, or at best suffering them and consenting to drag them along as a sort of necessitated incumbrance. Many a person's sexual life must be in this plight, and lead a kind of outlawed existence which would be intolerable save that once for all the eyes are closed upon it (St^e Beuve).[57]

It is interesting that James applies this analysis to sexuality, given that – unlike his contemporary Freud – he virtually never mentions this topic in his published writings. In this passage James makes specific reference to literary critic Charles Augustin Sainte-Beuve ('St^e Beuve'), who was known to have conducted an affair with the wife of Romantic poet Victor Hugo. However, his comments on the 'outlawed existence' of sexual lives are relevant in a broader sense. For instance, it has been clear at least since the work of twentieth-century American sex researcher Alfred Kinsey that a significant percentage of the population 'rounds off' its sexual identity to conform to societal expectations of exclusive heterosexuality or homosexuality.[58] Whether this counts as 'vicious' depends upon the level of rounding and one's attitude toward it. The self may legitimately be framed and guided by an abstract framework. Viciousness comes in when one denies the artifice involved in this process. One projects a two-dimensional image of oneself.

The reification of the homosexuality-heterosexuality dichotomy is intimately related to the reification of the male–female dichotomy. The traditional supposition is that one is either male or female, at which point one either exhibits the normal pattern of sexuality or is somehow inverted completely to its opposite. The dialectical interpenetration of these binaries in actual experience is a source for concern only from the perspective of a threatened self that seeks to enforce a false purity. One fears the 'more' that overflows one's identity, carrying the threat of instability or rejection. This is a problem for both individuals and cultures. Western science and philosophy have a pronounced masculinist streak that attempts to expel feminine-associated dimensions of experience, including emotion, passivity and non-conceptual awareness. This attitude is exemplified in the figure of the smug male scientist-in-training, who enters an adolescent epistemological culture while still an adolescent himself.

Such an attitude is a problem if, as Maslow contends, 'complete health means being available to yourself at all levels'.[59] Maslow's conception of health 'permits us to understand why psychologically healthy people are more able to enjoy, to love, to laugh, to have fun, to be humorous, to be silly, to be whimsical and fantastic, to be pleasantly "crazy"'.[60] A non-defensive attitude toward the unconscious and arational is allied with a generative conception of abstraction as a tool that makes experience tractable without presuming to supplant it.

The divided self

The above dialectical concepts may be brought to bear on a contentious question in the interpretation of William James. This is the question of James's *dividedness*. An examination of this question will serve as a useful coda to the present study, as it speaks to questions of self-transformation under a Pragmatic Image of Darwinism.

The theme of James's dividedness was thrust into the foreground by Richard Gale's provocative book *The Divided Self of William James* (1999). According to Gale, James is hopelessly divided in his conception of the individual and thus in his philosophy. On the one hand, James argues for a 'Promethean pragmatism' that construes the individual as active and purposive. In this mode, James views concepts instrumentally and relativizes ontological claims to purposes. On the other hand, James seems to argue (especially in *The Varieties of Religious Experience*) for an 'anti-Promethean mysticism' that construes the individual as passive and receptive. In this mode, James identifies meaning and reality with the direct apprehension of present reality. Both an action-oriented pragmatist and a mystic, James is ontologically trying to 'have it all':

> What James's quest to have it all most desires is to be both of these selves *at the same time*. What we really want is to be both a Sartrian *In-Itself* that self-sufficingly abides in its completeness within the present and a *For-Itself* that is always racing ahead of itself into the future so as to complete itself. In other words, we want to be God. Not surprisingly, this is forever beyond our grasp. To be human is to accept the unresolvable tension between wanting to be both at the same time. The best we can hope for is a taking-turns solution of the first-I'm-this-and-then-I'm-that sort. One does not solve this problem. One can only bear witness to it.[61]

Gale concludes that James's failure to reconcile these two visions amounts to his failure as a systematic philosopher. In Gale's metaphor, James is not so much a philosopher as an artist singing 'The Divided Self Blues'.

Of course, if James were 'merely' an artist – as he had studied to be as a teen-ager – then his writings could still be worthwhile as art. They are also *philosophical*, however, in the only sense in which James considered philosophy to matter: They imply a worldview according to which one may live. To see this requires taking a dialectical view of selfhood and of the role of the intellect in practical life.

As Gale recognizes, the tension in James's work is a constitutive feature of human existence: to desire both wholeness and constructive activity. Pure positive existence is an ideal resting place, but activity requires a negative space to open up between present and future. Such an emptiness threatens the wholeness of the present, but it must be posited and traversed to reach a new state of (provisional) wholeness. This happens in quotidian ways, as in the struggle to write a paragraph or negotiate a conversation with a friend. It also happens in bigger ways, as in the forging of a career path or the adoption of a metaphysical belief system. Dewey makes this point about life in general in *Art as Experience*:

> Life itself consists of phases in which the organism falls out of step with the march of surrounding things and then recovers unison with it – either through effort or by some happy chance. And, in a growing life, the recovery is never mere return to a prior state, for it is enriched by the state of disparity and resistance through which it has successfully passed.... Here in germ are balance and harmony attained through rhythm. Equilibrium comes about not mechanically and inertly but out of, and because of, tension.[62]

One is always falling out of step with one's situation and then finding a new way of being in step again. Things are not going wrong when experience exhibits such a tension between activity and passivity. *This is just what experience does.* The tension is irresolvable. Resolution is death.

Gale suggests that such a rhythm represents a 'taking-turns' solution, whereas we really desire both activity and passivity *at once*. However, such a desire is artificial, or in James's words 'intellectualistic'. One does not take turns from absolute wholeness to absolute emptiness. Such absolutes are fictions that are abstracted from experience. In reality, each 'turn' dialectically implies its other, which implies it back. In the final analysis, neither absolute activity nor absolute passivity – whether both-at-once or in turns – is comprehensible. To be 'God' in Gale's sense is not even a clear-eyed goal. What *is* comprehensible is the dynamogenic rhythm of activity and passivity that constitutes the conditions for finite freedom, meaning and creativity.

According to the present study, the relationship between activity and passivity takes a particular logical form on James's hierarchical selectionism: One is active insofar as one wilfully selects among possible thoughts and actions that construct the twofold habitual and social-natural environment; but one is also passive vis-à-vis these same possibilities in that one is incapable of producing what one selects. What is more, one may forego active willing entirely, as a show of trust toward a wider unconscious dimension of experience. Thus, different parts of oneself are active or passive toward one another in multiple senses of activity and passivity – and apparent passivity may be a kind of second-order activity in which periods of surrender are intentionally interwoven with a larger meaningful pattern. To reject such a model is not to swap the 'Divided Self Blues' for some up-beat pop number but to forego the kind of constrained and relative meaning that is possible.

In this context, James's central ethical question is when to exert one's will, as opposed to allowing unconscious forces the right of way. Crucially, James nowhere claims that overt action is better than inaction or that these two modes are mutually exclusive. Pragmatism does not mean frenetic or unintelligent action. To emphasize practice does not imply a particular way of weaving the relatively active into the relatively passive or of weighting their importance. It would not be un-pragmatist, for instance, to spend one's day meditating rather than working obsessively to complete a project of questionable value. In principle, there is no limit to what may count in the pragmatist's ledger as a satisfactory practical effect. James may stress activity and passivity differently in different texts, but he may be read charitably as integrating these phases of experience rather than prizing one over the other.[63]

Gale himself acknowledges in a response to his critics that it might not be a bad thing to be divided in the Jamesian sense.[64] The recalcitrant kernel of Gale's critique, however, is not that James gives us conflicting images of practical life. It is that these visions generate an awkward theoretical position. Namely, James seems to leave us with two distinct kinds of truth: conceptual truth that is judged in terms of its consequences; and non-conceptual truth through the direct apprehension of reality. James does impute significance to the ineffable light that peaks between the cracks of the conceptual edifice. If this light is understood, not merely as inspiration or spiritual innervation but as a direct kind of *knowledge*, then James seems to be proffering an 'error theory' on which the world-as-conceptualized does not track or represent a deeper world. On this view, we live day-by-day on pragmatic 'truth', while occasionally catching glimpses of intuitive Truth. Such a viewpoint may be troubling to those who equate reality with the rational or discussable – a commitment that shows up in different ways in Plato, Hegel and mechanistic science. It perhaps lies closer to certain Eastern views, such as the Buddhist doctrines of non-self and the impermanence of all perceived things.

An interesting feature of such a view is that it introduces a counterbalancing levity to the gravity of existence. A kind of liberation accompanies the idea that both self and world as habitually conceptualized are illusory. They are real but not *really real*. In many cases, the more gravely seriously one takes an endeavour – say, public speaking or a sexual act – the worse one may perform. Ironically, to deflate the ontological status of the conceptualized world may be a condition of negotiating it optimally, at least for agents of a certain temperament.

Of course, this is to read the mystical side of James through the Promethean side, by assessing mysticism in terms of its consequences for practice. Mysticism may be useful, but the mystical side of James wants his mystical insights also to be True. Perhaps mystical insight into the structure of reality may be True in a deep sense. Perhaps the very idea of such Truth is a vicious abstraction from the myriad truths that populate the work-a-day world. By its own lights, mysticism can be neither falsified nor corroborated by discursive means (as in this paragraph). About mysticism therefore nothing more will be said.

What can be said is that theory is a kind of practice birthed from processes that are just as arational and mysterious as mystical insight claims to be. We do not understand knowledge, whether intuitive or conceptual, by understanding its origins. We recognize it in practice, as one manifestation of a finite teleology constructed in real time in ontogeny, society and history. In bodying forth this teleology, each of us is both knower and doer; believer and doubter; thinker and feeler; self-identical and other; sensory perceiver and weaver of narratives; child of a time and maker of it. Instead of resting at one pole in these dualities, one gets elaborated – and in some measure elaborates oneself – through their interplay.

Notes

1 Lewontin 1991.
2 ECR, 321–326.
3 James is an existentialist in that he prescribes self-construction in the absence of a pre-given human essence or cosmic teleology. Notably, James provides a more constrained and naturalistic framework for self-transformation than does Jean-Paul Sartre's later doctrine of radical freedom. See also Bixler (1958) and Roth (1969).
4 MT, 134.
5 SPP, 56.

6 MT, 135–136. Similarly, James defines 'vicious intellectualism' as 'The treating of a name as excluding from the fact named what the name's definition fails positively to include' (PU, 32). To name something here is to categorize it, which is taken to exclude other possible categories.

7 MT, 136.

8 PU, 99.

9 Winther 2014.

10 Dawkins 1989/1976.

11 'The imploration is to stop the dichotomous thinking and pernicious reification of single models, and instead search for divisions of labor, complementarities, and legitimate redescriptions among the various extant models' (Winther 2011, 1).

12 James references William Wallace's translation of the 1830 expanded edition of Hegel's *Encyclopedia Logic* ('Shorter Logic').

13 According to Hegel (1991/1830), '[l]ogic is the science of the pure Idea', such that '[t]he Idea is thinking, not as formal thinking, but as the self-developing totality of its own peculiar determinations and laws, which thinking does not already *have* and find given within itself, but which it gives to itself' (§19).

14 Hegel 1991/1830, §19.

15 Hegel 1991/1830, §79. This triadic structure has a predecessor in the neo-Platonism of the third and fourth centuries AD. According to Proclus, for example, anything can be considered (1) in its permanence (by itself), (2) in its procession (emanating out of itself), and (3) in its reversion (returning back to itself). In Kant and Hegel these appear as logical moments, structuring Kant's table of categories and Hegel's speculative method. Thanks to Abraham Stone (personal communication).

16 Hegel 1991/1830, §81.

17 WB, 196.

18 PU, 45.

19 PU, 45.

20 PU, 52.

21 Hegel 1977/1807, §§90–110. This interpretation is in agreement with David Hoy (personal communication) and Schultz (2015), as against Morse (2005).

22 WB, 217.

23 WB, 221.

24 Levins and Lewontin 1985, 98.

25 For niche construction theory (a cousin of dialectical biology), see Odling-Smee *et al.* (2003) and Laland *et al.* (2003).

26 Oyama 2000, 31.

27 Oyama *et al.* 2003, 3–4. See also the related concept of the 'holobiont', a physiological and evolutionary unit defined by a multicellular eukaryote's relationship to its persistent symbionts (Chiu and Gilbert 2015).

28 Dennett 1975.

29 Maturana and Varela 1972. See also Varela *et al.* (1991) and Thompson (2007).

30 Varela 1991, 81.

31 See also Pradeu (2010).

32 Varela 1991, 80.

33 For a challenge to that reading of Kant, see Weber and Varela (2002).

34 A major exception here is Eugene Taylor. See Taylor (1991) and subsequent references.

35 Taylor 1984; 1996.

36 McDermott 1976.

37 ERE, 42.

38 On James and Allport in particular, see High and Woodward (1980).

39 The study of human flourishing within psychology has enjoyed something of a comeback in the positive psychology movement, launched by Martin Seligman at

the turn of the twenty-first century. However, the latter movement made its name in part by distinguishing itself from the 'unscientific' qualities of humanistic psychology. This caused a rift and opened positive psychology to the charge that it is *positivist* psychology. See Seligman and Csikzentmihalyi (2000) and Taylor (2001). For James and positive psychology, see Pawelski (2003a).

40 Taylor (2010).
41 This resembles Polish psychologist Kazimierz Dabrowski's (1970) theory of 'positive disintegration'. See Dabrowski *et al.* (1970).
42 Maslow 1971, 218.
43 A note of James's from 1869 shows a similar awareness of the dialectic of growth and fear: 'The expansive, embracing tendency, the centripetal, defensive, forming two different modes of self-assertion: sympathy and self-sufficingness. (The two combine and give respect?) To "accept the universe", to protest against it, *voluntary* alternatives'. Quoted in Perry (1935 I, 301–302).
44 Maslow 1971, 24.
45 TT, 113–114.
46 Maslow 1971, 43–47.
47 Maslow's peak experience is a clear ancestor of Csikzentmihalyi's (1990) concept of 'flow'.
48 Maslow illustrates this point with 'internationalist member of the human species' William James (Maslow 1968, 181n2).
49 VRE, 298–300. See Chapter 4 of the present study.
50 As Maslow points out, this book pairs well with Michael Polanyi's *Personal Knowledge* (1958).
51 If Nietzsche is to be believed, science even serves an ascetic need to face oneself with difficult truths. See Chapter 4 of the present study.
52 'I suppose it is tempting, if the only tool you have is a hammer, to treat everything as if it were a nail' (Maslow 1966, 16). Another source for this adage is Abraham Kaplan's *The Conduct of Inquiry* (1964).
53 Maslow 1966, 13–16.
54 For example, see Maslow (1971, 155).
55 On mapping and pluralism in science, see Winther (in preparation).
56 For instance, see Hacking (1995).
57 MEN, 314–315.
58 Kinsey *et al.* 1948; Kinsey *et al.* 1953.
59 Maslow 1971, 88. This is not to claim that Maslow himself radically critiqued concepts of sexuality, sex or gender. He was somewhat conservative in this regard, at least by twenty-first century standards.
60 Maslow 1968, 209.
61 Gale 1999, 332.
62 Dewey 1934, 12–13.
63 For instance, Pawelski (2007) provides a 'developmental' reading on which James developed an integrative view of activity and passivity toward the end of his life.
64 Gale 2004. Gale is responding to Pawelski (2003b) and Cooper (2002).

References

Bixler, J. S. 1958. The Existentialists and William James. *The American Scholar* 80–90.
Chiu, L. and S. F. Gilbert. 2015. The Birth of the Holobiont: Multi-species Birthing Through Mutual Scaffolding and Niche Construction. *Biosemiotics* 8, no. 2: 1–20.
Cooper, W. 2002. *The Unity of William James's Thought*. Nashville: Vanderbilt University Press.

Csikszentmihalyi, M. 1990. *Flow: The Psychology of Optimal Experience*. New York: Harper and Row.

Dabrowski, K., A. Kawczak and M. M. Piechowski. 1970. *Mental Growth through Positive Disintegration*. London: Gryf Publications.

Darwin, C. 1859. *On the Origin of Species By Means of Natural Selection, Or The Preservation of Favoured Races in the Struggle for Life*. London: John Murray.

Dawkins, R. 1989. *The Selfish Gene*. 2nd edn. Oxford: Oxford University Press. Original edition, 1976.

Dennett, D. C. 1975. Why the Law of Effect Will Not Go Away. *Journal for the Theory of Social Behaviour* 5, no. 2: 169–188.

Dewey, J. 1934. *Art as Experience*. New York: Minton, Balch & Co.

Gale, R. M. 1999. *The Divided Self of William James*. Cambridge: Cambridge University Press.

Gale, R. M. 2004. The Still-Divided Self of William James: A Response to Pawelski and Cooper. *Transactions of the Charles S. Peirce Society* 40, no. 1: 153–170.

Hacking, I. 1995. *Rewriting the Soul: Multiple Personality and the Sciences of Memory*. Princeton, NJ: Princeton University Press.

Hegel, G. W. F. 1873. *Hegel's Logic: Being Part One of the Encyclopedia of the Philosophical Sciences*. Translated by W. Wallace. New York: Oxford University Press. Original edition, 1830.

Hegel, G. W. F. 1977. *The Phenomenology of Spirit*. 5th edn. Translated by A. V. Miller. Oxford: Oxford University Press. Original edition, 1807.

Hegel, G. W. F. 1991. *The Encyclopaedia Logic, with the Zusätze: Part I of the Encyclopaedia of Philosophical Sciences*. 3rd edn. Translated by T. F. Geraets, W. A. Suchting and H. S. Harris. Edited by T. F. Geraets, W. A. Suchting and H. S. Harris. Indianapolis, IN: Hackett. Original edition, 1830.

High, R. P. and W. R. Woodward. 1980. William James and Gordon Allport: Parallels in Their Maturing Conceptions of Self and Personality. In *Psychology, Theoretical-historical Perspectives: Theoretical-historical Perspectives*, eds R. W. Rieber and K. Salzinger, 57–79. New York: Academic Press.

James, W. 1975. *The Meaning of Truth*. The Works of William James. Edited by F. Burkhardt, F. Bowers and I. K. Skrupskelis. Cambridge, MA: Harvard University Press. Original edition, 1909.

James, W. 1976. *Essays in Radical Empiricism*. The Works of William James. Edited by F. Burkhardt, F. Bowers and I. K. Skrupskelis. Cambridge, MA: Harvard University Press. Original edition, 1912.

James, W. 1977. *A Pluralistic Universe*. The Works of William James. Edited by F. Burkhardt, F. Bowers and I. K. Skrupskelis. Cambridge, MA: Harvard University Press. Original edition, 1908.

James, W. 1979. *Some Problems of Philosophy*. The Works of William James. Edited by F. Burkhardt, F. Bowers and I. K. Skrupskelis. Cambridge, MA: Harvard University Press. Original edition, 1911.

James, W. 1979. *The Will to Believe and Other Essays in Popular Philosophy*. The Works of William James. Edited by F. Burkhardt, F. Bowers and I. K. Skrupskelis. Cambridge, MA: Harvard University Press. Original edition, 1897.

James, W. 1981. *The Principles of Psychology*. 2 vols. The Works of William James. Edited by F. Burkhardt, F. Bowers and I. K. Skrupskelis. Cambridge, MA: Harvard University Press. Original edition, 1890.

James, W. 1983. *Talks to Teachers on Psychology: And to Students on Some of Life's*

Ideals. Edited by F. Burkhardt, F. Bowers and I. K. Skrupskelis. Cambridge, MA: Harvard University Press. Original edition, 1899.

James, W. 1985. *The Varieties of Religious Experience*. The Works of William James. Edited by F. Burkhardt, F. Bowers and I. K. Skrupskelis. Cambridge, MA: Harvard University Press. Original edition, 1902.

James, W. 1987. *Essays, Comments, and Reviews*. The Works of William James. Edited by F. Burkhardt, F. Bowers and I. K. Skrupskelis. Cambridge, MA: Harvard University Press.

James, W. 1988. *Manuscript Essays and Notes*. The Works of William James. Edited by F. Burkhardt, F. Bowers and I. K. Skrupskelis. Cambridge, MA: Harvard University Press.

Kant, I. 1914. *Critique of Judgment*. Translated by J. H. Bernard. London: Macmillan. Original edition, 1790.

Kant, I. 1998. *Critique of Pure Reason*. Translated by P. G. Guyer and A. W. Wood. Cambridge: Cambridge University Press. Original edition, 1781.

Kaplan, A. 1964. *The Conduct of Inquiry: Methodology for Behavioral Science*. San Francisco, CA: Chandler Publishing Company.

Kinsey, A. C., W. B. Pomeroy and C. E. Martin. 1948. *Sexual Behavior in the Human Male*. Philadelphia, PA: W.B. Saunders.

Kinsey, A. C., W. B. Pomeroy, C. E. Martin and P. H. Gebhard. 1953. *Sexual Behavior in the Human Female*. Philadelphia, PA: W.B. Saunders.

Laland, K. N., F. J. Odling-Smee and M. W. Feldman. 2003. Niche Construction, Ecological Inheritance, and Cycles of Contingency in Evolution. In *Cycles of Contingency: Developmental Systems and Evolution*, eds S. Oyama, P. E. Griffiths and R. D. Gray, 117–126. Cambridge, MA: The MIT Press.

Levins, R. and R. C. Lewontin. 1985. *The Dialectal Biologist*. Cambridge, MA: Harvard University Press.

Lewontin, R. C. 1991. Foreword. In *Organism and the Origins of Self*, xiii-xix. Dordrecht, The Netherlands: Kluwer Academic Publishers.

Maslow, A. H. 1966. *The Psychology of Science: A Reconnaissance*. New York: Harper & Row.

Maslow, A. H. 1968. *Toward a Psychology of Being*. 2nd edn. New York: Van Nostrand Reinhold.

Maslow, A. H. 1971. *The Farther Reaches of Human Nature*. New York: Viking Press.

Maturana, H. R. and F. J. Varela. 1972. *De Máquinas y Seres Vivos*. Santiago, Chile: Editorial Universitaria.

McDermott, J. J. 1976. *The Culture of Experience: Philosophical Essays in the American Grain*. New York: New York University Press.

Morse, D. 2005. William James's Neglected Critique of Hegel. *Idealistic Studies* 35, no. 2–3: 199–213.

Odling-Smee, F. J., K. N. Laland and M. W. Feldman. 2003. *Niche Construction: The Neglected Process in Evolution*. Princeton, NJ: Princeton University Press.

Oyama, S. 2000. *The Ontogeny of Information: Developmental Systems and Evolution*. 2nd edn. Science and Cultural Theory. Edited by B. H. Smith and R. E. Weintraub. Durham, NC: Duke University Press. Original edition, 1985.

Oyama, S., P. E. Griffiths and R. D. Gray. 2003. *Cycles of Contingency: Developmental Systems and Evolution*. Cambridge, MA: MIT Press.

Pawelski, J. O. 2003a. William James, Positive Psychology, and Healthy-Mindedness. *The Journal of Speculative Philosophy* 17, no. 1: 53–67.

Pawelski, J. O. 2003b. William James's Divided Self and the Process of Its Unification: A Reply to Richard Gale. *Transactions of the Charles S. Peirce Society* 39, no. 4: 645–656.

Pawelski, J. O. 2007. *The Dynamic Individualism of William James*. Albany, NY: State University of New York Press.

Perry, R. B. 1935. *The Thought and Character of William James*. 2 vols. Boston, MA: Little, Brown, and Company.

Polanyi, M. 1958. *Personal Knowledge: Towards a Post-Critical Philosophy*. New York: Harper and Row.

Pradeu, T. 2010. What is an Organism? An Immunological Answer. *History and Philosophy of the Life sciences* no. 32: 247–267.

Roth, J. K. 1969. *Freedom and the Moral Life: The Ethics of William James*. Philadelphia: Westminster Press.

Schultz, L. C. 2015. Pluralism and Dialectic: On James's Relation to Hegel. *Hegel Bulletin* 36, no. 2: 202–224.

Seligman, M. E. P. and M. Csikszentmihalyi. 2014. Positive Psychology: An Introduction. *American Psychologist* 55, no. 1: 5–14.

Taylor, E. 1984. *William James on Exceptional Mental States: The 1896 Lowell lectures*. New York: Scribner.

Taylor, E. 1991. William James and the Humanistic Tradition. *Journal of Humanistic Psychology* 31, no. 1: 56–74.

Taylor, E. 1996. *William James on Consciousness beyond the Margin*. Princeton, NJ: Princeton University Press.

Taylor, E. 2001. Positive Psychology and Humanistic Psychology: A Reply to Seligman. *Journal of Humanistic Psychology* 41, no. 1: 13–29.

Taylor, E. 2010. William James and the Humanistic Implications of the Neuroscience Revolution: An Outrageous Hypothesis. *Journal of Humanistic Psychology* 50, no. 4: 410–429.

Thompson, E. 2007. *Mind in Life: Biology, Phenomenology, and the Sciences of Mind*. Cambridge, MA: Harvard University Press.

Varela, F. J. 1991. Organism: A Meshwork of Selfless Selves. In *Organism and the Origins of Self*, ed. A. I. Tauber, 79–107. Dordrecht, The Netherlands: Kluwer Academic Publishers.

Varela, F. J., E. Thompson and E. Rosch. 1991. *The Embodied Mind: Cognitive Science and Human Experience*. Cambridge, MA: MIT Press.

Weber, A., and F. J. Varela. 2002. Life after Kant: Natural Purposes and the Autopoietic Foundations of Biological Individuality. *Phenomenology and the Cognitive Sciences* 1, no. 2: 97–215.

Winther, R. G. 2011. Consciousness Modeled: Reification and Promising Pluralism. *Pensamiento* 67, no. 254: 617–630.

Winther, R. G. 2014. James and Dewey on Abstraction. *The Pluralist* 9, no. 2: 1–28.

Winther, R. G. In preparation. *When Maps Become the World*. Chicago, IL: University of Chicago Press.

Index

Page numbers in **bold** denote figures.

Lehrman, Daniel 61
Levins, Richard 66, 163
L'évolution créatrice (Bergson) 145
Lewontin, Richard 48, 66, 138, 156, 163, 164
libertarianism 83
Literary Remains (James, Snr.) 141
Little Book of Life After Death (Fechner) 147
Locke, John 18
logic 146, 157, 160–2; *see also* evolutionary logics
Longino, Helen 139, 140
Lorenz, Konrad 61
Lutoslawski, Wincenty 147

McDermott, John 12, 167
McDougall, William 61
Malthus, Thomas 3, 36, 56–7
'Many and the One, The' (James) 141
Maslow, Abraham 168–9, 170
Materials for the Study of Variation (Bateson) 48–9
Maturana, Humberto 166
Mayr, Ernst 3
meaning 133, 136–8, 149–50
Meaning of Truth, The (James) 9, 116, 132, 138, 143
mechanism 15, 22n11; of biological adaptation 38; of heredity 40; of the inheritance of acquired characteristics 30, 38; mechanistic determinism 4, 129, 156; mechanistic science 4, 103–4, 174; of natural selection 30; new mechanism school 22n11; non-purposive sorting mechanism 47
meme theory 62–4
Metaphysical Club 33–4, 143, 157
metaphysics 45, 106, 121, 140, 141, 143, 145, 150, 167; *see also* radical empiricism
mind 38–9, 44, 61, 74, 143, 165; structural unit of 75–6
monism 141–2
Moore, G. E. 129
Moore, Gregory 106
moral equivalent of war concept 9, 117
'Moral Philosopher, The, and the Moral Life' (essay, James) 11, 16, 104, 111, 116, 130–2, 138–9, 147
moral psychology 9, 11, 15, 16, 21, 78, 79–80, 84–5, 87, 111; energy and will-to-power as moral fuel 104–6
Morgan, Conwy Lloyd 61

Moss, Lenny 17
Münsterberg, Hugo 9, 41, 88
Murray, Henry 168
Mussolini, Benito 121
mutation theory 37; *see also* saltationism
mysticism 16, 72, 141, 146, 172, 174

National Socialism 121
natural selection 1–2, 5, 8, 20, 34, 43, 44–5, 91, 156, 157; and essentialism 2–3; externalist construal of 46; and the integrated theory of development, heredity and variation 35–6; as a mindless and non-intentional process 2; and probabilistic evidence 3; units of selection controversy 15; and variation 78–9; *see also* selectionism
naturalized apriorism 43
Nervous system 40, 49, 78, 80, 81; sensorimotor system 75, 76, 79, 166; *see also* reflex arc
Nietzsche, Friedrich 17, 21, 80, 137, 156, 158; character ideals and evolutionary logics, comparison with James 97–128; James's view of Nietzsche as 'dying rat' 100–4; similarities in James's and Nietzsche's lives 97–9
Nordau, Max 99
normativity 137, 140
'nothing but' philosophy 4, 156, 159
novelty 83, 92, 142–3, 145

objectivity 139–40
objects, perception of 77–8
'On a Certain Blindness in Human Beings' (essay, James) 87–8, 117
'On Exceptional Mental States' (lectures, James) 9, 167
'On Some Omissions of Introspective Psychology' (essay, James) 143
On the Classification of Animals, and on the Vertebrate Skull (Huxley) 34
On the Genealogy of Morals (Nietzsche) 98, 107
On the Origin of Species (Darwin) 1, 7, 14, 37, 73, 157
ontogeny 5, 15, 30, 39, 50, 82, 101, 108, 140–1, 147, 150, 157, 164; mental ontogeny 48–9; social evolution as ontogeny 65–6
Ontogeny of Information, The (Oyama) 140–1, 163–4
'Origin of Human Races, The' (essay, Wallace) 35